Protein Refolding

ACS SYMPOSIUM SERIES **470**

Protein Refolding

George Georgiou, EDITOR
University of Texas

Eliana De Bernardez-Clark, EDITOR
Tufts University

Developed from a symposium sponsored
by the Divisions of Biochemical Technology
and Biological Chemistry
at the 199th National Meeting
of the American Chemical Society,
Boston, Massachusetts,
April 22–27, 1990

American Chemical Society, Washington, DC 1991

SEP/AE
CHEM

Library of Congress Cataloging-in-Publication Data

Protein refolding / George Georgiou, editor, Eliana De Bernardez-Clark, editor.

p. cm.—(ACS symposium series; 470).

"Developed from a symposium sponsored by the Divisions of Biochemical Technology and Biological Chemistry at the 199th National Meeting of the American Chemical Society, Boston, Massachusetts, April 22–27, 1990."

Includes bibliographical references and index.

ISBN 0–8412–2107–3

1. Protein folding—Congresses.

I. Georgiou, George. II. De Bernardez-Clark, Eliana. III. American Chemical Society. Division of Biochemical Technology. IV. American Chemical Society. Division of Biological Chemistry. V. American Chemical Society. Meeting (199th: 1990: Boston, Mass.) VI. Series.

QP550.P76 1991   91–22163
547.7′5—dc20   CIP

SD 9/9/91 QA

ACS Symposium Series

QP550
P76
1991
CHEM

M. Joan Comstock, *Series Editor*

1991 ACS Books Advisory Board

Foreword

THE ACS SYMPOSIUM SERIES was founded in 1974 to provide a medium for publishing symposia quickly in book form. The format of the Series parallels that of the continuing ADVANCES IN CHEMISTRY SERIES except that, in order to save time, the papers are not typeset, but are reproduced as they are submitted by the authors in camera-ready form. Papers are reviewed under the supervision of the editors with the assistance of the Advisory Board and are selected to maintain the integrity of the symposia. Both reviews and reports of research are acceptable, because symposia may embrace both types of presentation. However, verbatim reproductions of previously published papers are not accepted.

Contents

Preface

THE AMINO ACID SEQUENCE OF A PROTEIN determines its three-dimensional structure, as was shown 30 years ago by Afinsen and his co-workers. However, understanding how a protein folds to its native conformation still represents a major challenge to scientists. Protein folding is also a significant problem in biotechnology. The production of genetically engineered polypeptides frequently results in the accumulation of insoluble protein aggregates within the cells. These aggregates, known as inclusion bodies, must be solubilized and refolded to produce a biologically active product. The refolding process is often the critical bottleneck in the production of high-value proteins.

Recently, our understanding of the forces governing protein folding has greatly improved because of carefully designed detailed studies of the folding of small globular proteins. However, some of these successes remain to be translated into technological advantages.

This book attempts to bridge the gap between fundamental and applied studies on protein folding. Some of the issues addressed include in vivo protein folding, protein aggregation and inclusion body formation, elucidation of the folding pathway, characterization of folding intermediates, and practical considerations in protein renaturation.

We gratefully acknowledge the assistance of many reviewers who provided the authors with helpful comments and suggestions. We would also like to thank the ACS Books Department for their help throughout the process of editing this book.

GEORGE GEORGIOU
University of Texas
Austin, TX 78712

ELIANA DE BERNARDEZ-CLARK
Tufts University
Medford, MA 02155

March 7, 1991

Chapter 1

Inclusion Bodies and Recovery of Proteins from the Aggregated State

Eliana De Bernardez-Clark[1] and George Georgiou[2]

[1]Department of Chemical Engineering, Tufts University,
Medford, MA 02155
[2]Department of Chemical Engineering, University of Texas—Austin,
Austin, TX 78712

In recent years protein folding has emerged as a central issue in cell physiology, immunology and biotechnology. *In-vivo* protein folding plays an important role in cellular responses such as the transport of polypeptides through membranes, cellular adaptation to extreme conditions, the assembly of viruses and multimeric enzymes and the destruction of aberrant polypeptides by proteases. Folding also appears to be directly related to certain disease states and immunosupression (*1,2*).

In biotechnology the synthesis of biologically active recombinant proteins is the key issue for the production of commercially important polypeptides and when the expression of foreign genes is employed to manipulate the metabolism of the cell. Frequently, recombinant polypeptides are not able to fold properly within the host cell and form amorphous protein aggregates usually consisting of misfolded, often denatured polypeptides. These aggregates are known as inclusion bodies or refractile bodies since they appear as highly refractile areas when the cells are observed microscopically (*3,4*). The aggregation of recombinant proteins was first observed with human insulin expressed in *Escherichia coli*. However, it is now well established that inclusion bodies are formed in other gram-negative as well as in gram-positive bacteria (*5,6*). Furthermore, several recombinant proteins have been shown to accumulate in insoluble form in *Saccharomyces cerevisiae*, other lower eucaryots, insect cells and even animal cells (*7,8*). Aggregation is particularly important for the development and production of engineered proteins. Mutants (often called muteins), derived by amino acid substitutions of naturally occurring proteins and *ab-initio* designed polypeptides are frequently unable to fold properly within the cell. The folding pathways for such proteins can be different from their naturally occurring homologues. For this reason, it is often very difficult to recover biologically active proteins by conventional procedures, *i.e.* by solubilization of the intracellular aggregates *in-vitro* and renaturation.

The formation of inclusion bodies is by no means an aberrant phenomenon related only to the expression of heterologous polypeptides. Protein aggregation *in-vivo* is also observed with native proteins in cells carrying certain genetic lesions or grown in stressful environments. Noteworthy examples of the latter include the accumulation of ß-galactosidase in insoluble form in *Escherichia coli* cells grown in the presence of the amino acid pyromycin (*9*), and the formation of protein aggregates in cells subjected to heat shock (*10*). Thus, the formation of inclusion bodies of recombinant proteins is just one of the manifestations of protein misfolding. As will be discussed later, aggregation results from the accumulation of partially folded polypeptide chains that

0097–6156/91/0470–0001$06.00/0

fail to reach the native conformation. The chapters by Mitraki *et al*. and Chen and King illustrate how the study of aggregation can provide important information on the mechanism of polypeptide folding within the cell.

In biotechnology, the need to recover functional polypeptides from aggregated material has forced a reevaluation of conventional protein production and purification strategies. Normally, soluble protein products are purified from other cellular components by a series of chromatographic separations conducted under non-denaturing conditions (*11*). The objective is to remove impurities while recovering the highest possible amount of active material. On the other hand, aggregated polypeptides are usually misfolded and therefore devoid of any biological activity. The key issue in product recovery is to reconstitute the correct three-dimensional conformation. In some cases resuspension of the inclusion bodies in buffers of high ionic strengths or in the presence of ion exchange resins is sufficient to solubilize the protein in an active conformation (*12,13*). In a recent study it was observed that protein aggregates of human apolipoprotein formed in the periplasmic space of *E.coli* could be solubilized by dilution (*14*). As a rule however, aggregated polypeptides must be first solubilized under denaturing conditions such as concentrated solutions of chaotropic agents, detergents, extreme pH or organic solvents. The yield of active protein following renaturation from the solubilized state can vary greatly depending on the amino acid sequence, refolding conditions and the presence of impurities (originating from the inclusion bodies).

When a protein can be refolded at a high yield by using a straightforward procedure, then the formation of inclusion bodies is an advantage rather than an impediment for protein production. Soluble heterologous proteins are often susceptible to extensive degradation within the host cell (*15*). Proteolysis results in low yields of the desired products despite high rates of transcription and translation. Cheng *et al*. (*16*) have reported that intracellular proteases do not attack inclusion body proteins thereby increasing the stability of overexpressed proteins. Aggregation prevents degradation, presumably because protease sensitive sites become inaccessible (*15*). Not only will protein accumulation be greater in such systems, but losses due to proteolytic clipping will also be reduced. Clipped proteins, which are large fragments of a given protein that are not biologically active, often prove difficult to remove and act to complicate purification schemes. In fact, certain foreign polypeptides cannot be produced in intact form unless they are expressed in an aggregated, protease-resistant conformation. For this reason, proteins that are normally soluble within the host cell are sometimes purposely engineered to form aggregates (*7*). However, there are some instances where proteins in inclusion bodies are susceptible to degradation. Babbitt *et al*. (*17* and this volume) showed that an *E.coli* proteolytic enzyme co-aggregates with creatine kinase inclusion bodies. This enzyme is activated upon refolding leading to extensive degradation. Bowden and Georgiou have found that TEM ß-lactamase inclusion bodies formed under different expression conditions exhibit significant differences with respect to their resistance to trypsin *in-vitro* (unpublished observations). These observations indicate that the expression of polypeptides within protein aggregates does not always guarantee protection from degradation.

The formation of inclusion bodies also offers distinct advantages for the production of proteins that are toxic to the host cell and cannot be expressed in active form. Finally, protein aggregation allows to monitor the productivity of recombinant bacterial cultures on-line. Intracellular protein particles are highly refractile and affect the light scattering properties of the cell (*18*). The light scattering distribution obtained from a cell sorting apparatus may be used for fermentation monitoring and control. In contrast, the amount of soluble proteins must be determined by time consuming procedures such as enzymatic assays or gel electrophoresis.

The accumulation of recombinant proteins in an insoluble form offers even more significant benefits for downstream processing. Typically, the desired protein product represents at least 40 to 50 % of the total polypeptide content of inclusion bodies. The protein aggregates can be easily separated from the soluble components of lysed cells by centrifugation or microfiltration. Non-proteinaceous impurities such as lipids, ribosomes and nucleic acids represent a small fraction of the nominal mass of the inclusion bodies (Valax and Georgiou, unpublished observations). Depending on the separation procedure, membranes and nucleic acids sometimes cofractionate with the protein particles and appear to be present in larger amounts. In any case, these impurities can be easily extracted from the inclusion bodies to provide a highly purified form of the product for subsequent steps. The aggregated protein is then solubilized under denaturing conditions. The product may be purified from residual polypeptide impurities in the denatured state by employing high resolution separation techniques such as hydrophobic or reverse phase chromatography. Finally the purified protein is renatured under conditions that ensure the highest possible yield of the biologically active species.

It must be emphasized that the the expression of proteins in insoluble form is advantageous only when the recovery of the active product is sufficiently high. Currently several commercially important polypeptides including human insulin, somatotropins and interleukins are produced from *Escherichia coli* inclusion bodies. Although these are relatively small and simple molecules, there is increasing evidence that even complex monomeric proteins can be refolded *in-vitro* with reasonable efficiency. A very impressive example is the production of biologically active tissue plasminogen activator (TPA) from *Escherichia coli* inclusion bodies. The folding pathway for this monomeric protein is very complicated and depends on the correct pairing and oxidation of 34 out of the 35 cysteines of the molecule into 17 disulfide bonds. Nevertheless, high yields of active TPA at appreciable final protein concentrations (>100 µg/ml) have been obtained through the use of optimal conditions and an ingenious fed-batch refolding procedure (*19*). However, these impressive success stories should not obscure the fact that protein refolding remains a challenge. Complex proteins requiring the assembly of many subunits or the incorporation of prosthetic groups can rarely be reconstituted *in-vitro*.

In summary, protein folding is presently a very significant issue in both fundamental and applied biological research. Progress in the efficient production of recombinant proteins depends closely on the understanding of the physicochemical interactions, genetics and biochemistry of the processes which are responsible for the generation of the correct three dimensional structure. This chapter is intended to highlight the key aspects of protein folding, aggregation and refolding both in the laboratory and in industry.

Protein folding thermodynamics and kinetics

The amino acid interactions that determine the formation of stable secondary and tertiary structures can only be characterized in a well defined chemical environment. For this reason, the energetic and kinetic aspects of protein folding have been deduced primarily from *in-vitro* experiments.

The three dimensional structure of proteins in aqueous solutions is stable within a narrow range of temperatures, pH and cosolvent concentrations. Under these conditions the folded state is generally (but not universally) considered to be energetically favored, *i.e.* its Gibbs free energy is lower than any other possible conformation (*20*). A free energy minimum implies that the correctly folded protein is in equilibrium with molecules in a non-native conformation. Relatively small perturbations in temperature or solvent conditions shift the equilibrium towards the non-native species. The folding transition is a highly cooperative process. For small proteins, the change in the fraction of native protein as a function of temperature or denaturant concentration can be described well by a two state model. If the two state model applies, partially folded species missing some of the native interactions of the native protein should not be detectable at equilibrium. In contrast, larger proteins often consist of structural domains which can form independently of each other. Species with a partially folded structure can be distinguished by spectroscopic techniques and can be populated (*i.e.* their relative concentration increased) at moderately denaturing conditions.

The change in free energy that accompanies the transition between the unfolded and the folded state of the polypeptide is generally small, typically between 20-45 KJ/mol. This free energy difference provides the driving force for folding. However, the transition from the unfolded state which approximates a random coil to an intricate three-dimensional structure, is a very complex process. The initial events in folding have not yet been clearly defined (*21*). It appears that folding begins with the formation of certain elements of secondary structure involving neighboring amino acids or the collapse of the polypeptide chain into a conformation that minimizes the exposure of hydrophobic amino acids to the solvent (the molten globule state). The molten globule model is supported by the results of recent experiments which have provided detailed information regarding the structure of folding intermediates (*22*). Specific interactions between distant parts of the molecule may serve to stabilize the folded structure. Kim and colleagues have demonstrated that the formation of secondary structure in different peptide segments of the bovine pancreatic trypsin inhibitor (BPTI) cannot occur unless the peptides are joined via the appropriate disulfide bonds (*23,24*).

The rate of folding of different proteins can vary from less than one second to several hours and even days. Some polypeptides fold by two (or more) pathways which exhibit distinctly different kinetics (*25*). As a rule, folding is slow when it involves intermolecular interactions (*e.g.* assembly of polypeptide chains and prosthetic groups), rate limiting isomerization steps around peptide bonds or finally, covalent modification(s) of the polypeptide chain. When the rate of formation of a partially folded molecule exceeds the rate at which the subsequent rearrangement steps take place, then a *kinetic* intermediate accumulates at a detectable level. Depending on the pH, temperature and denaturant concentration, the dissipation of such intermediates may take from seconds to hours. The folding pathway of many proteins involves the formation of several intermediates. Extensive studies by Creighton and others have illustrated the complex pattern of folding intermediates formed during the refolding of BPTI (*20, 24*).

Kinetic intermediates can be identified by spectroscopic techniques or chemical trapping. However, their structural characterization has proved very difficult. Folding intermediates have an increased tendency to aggregate and therefore it is not possible to obtain the relatively concentrated solutions required for high resolution structural analysis by two-dimensional NMR. The problems related to structural studies of associated and aggregated proteins are discussed in a subsequent chapter by Katherine Schein. Recently a very powerful technique was developed for monitoring the folding rates of different parts of the polypeptide chain. Detailed structural characterization of partially folded intermediates of small proteins such as ribonuclease, cytochrome c, apomyoglobin, barnase and lysozyme was obtained by trapping slowly exchanging protons and analyzing them using 2-D NMR (*26-28*, see also Elöve and Roder, this volume, for a detailed discussion).

Protein aggregation is an important problem in renaturation studies. *In-vitro* folding and aggregation appear to be parallel processes (Mitraki *et al.*, this volume). For monomeric proteins, folding is a unimolecular event exhibiting a first order dependency on the polypeptide concentration. In contrast, aggregation involves the interaction of at least two polypeptide chains and follows higher order apparent kinetics (*29,30*). Consequently, aggregation is favored at elevated protein concentrations. The relation between protein concentration and self association is particularly important in the context of folding within the cell.

Intracellular Folding Events

The folding of proteins in the cell proceeds under conditions markedly different from those employed in refolding studies. Although little conclusive information is available it is generally thought that the *in-vivo* folding pathway is directly affected by the synthesis, covalent modification and localization of the polypeptide chain.

The issue of whether the rate of translation is important for folding has been contested for at least two decades. Completion of the nascent polypeptide chain during ribosomal synthesis can take from seconds to several minutes depending on the size of the protein, the organism and the growth rate. Neighboring amino acids in the incomplete chain are likely to be able to participate in the initial events of folding such as the formation of certain elements of secondary structure (*31*). However, the relation between protein synthesis and folding is uncertain. There is at least one protein, the enzyme phosphoribosyl anthranilate isomerase from *E.coli*, for which it has been shown that translation is not coupled to folding (*32*).

The localization of a polypeptide chain within a non-cytoplasmic cellular location can affect folding in a variety of ways. For steric and energetic reasons secreted proteins must be unfolded during transport so that they can snake through the lipid bilayer in a more or less vectorial fashion. The premature folding of newly synthesized secreted proteins in the cytoplasm is prevented by the signal sequence and through interactions with cellular factors such as chaperonins. The native conformation is obtained only after export has been completed. As may be expected, the vectorial transport of the amino acid chain through the membrane and the environment of secretory compartments exert a profound effect on folding.

The removal of the signal sequence which directs the transport of secreted proteins is a critical step in protein export. Extensive studies by Randall, Hardy and coworkers have demonstrated that the leader sequence prevents the proper folding of the mature polypeptide within the cytoplasm (*33,34*). For the *E.coli* ß-lactamase, the presence of the leader sequence decelerates the folding kinetics but does not prevent the formation of the enzymatically active conformation (*35*). Bowden and Georgiou (*36*) showed that the mode of translocation of ß-lactamase across the cytoplasmic membrane of *Escherichia coli* exerts a profound effect on the folding of the mature protein following secretion, presumably by affecting the unfolded state in the periplasmic space.

The conditions in secretory compartments, for example pH, redox potential and macromolecular composition are considerably different from those of the cytoplasm. Secreted proteins are also subject to a variety of chemical modifications such as acylation, glycosylation, the removal of signal or leader peptides, and formation of disulfide bonds. Protein glycosylation which occurs in the endoplasmic reticulum of eucaryotic cells has been shown to affect protein solubility and folding (*37*). Even more importantly, the pairing and oxidation of cysteines to form disulfide bonds occurs primarily at non-cytoplasmic locations which have the proper oxidizing environment. Clearly, the temporal and topological sequence of post-translational modifications of the polypeptide chain must exert a significant role on the folding pathway.

With the exception of disulfide bonds, all post-translational modifications must be catalyzed by cellular enzymes. The formation of disulfide bonds can occur at appreciable rates in the absence of enzymes and involves two steps: (i) the relatively rapid pairing of cysteines to form S-S bonds that do not correspond to those in the native structure and (ii) disulfide rearrangement (*20*). The isomerization of disulfide bonds to form the correct cysteine pairs that are present in the native protein is slow and represents an important rate limiting step in folding. For this reason the *in-vitro* refolding of polypeptides containing several cysteines is usually very slow and inefficient. In eucaryotic cells the formation of the correct disulfide bonds is accelerated by the enzyme protein disulfide isomerase or PDI (*38,39*). PDI is located in the luminar side of the endoplasmic reticulum where the environment favors cysteine oxidation. It is a very abundant enzyme and represents up to 2% of the total protein of certain mammalian cells. At present, there is no clear evidence of a disulfide isomerase activity in bacteria. The absence of such an enzyme from *E. coli* may be partly responsible for the misfolding of heterologous proteins with many cysteines.

Other than the formation of disulfide bonds, the only well characterized rate limiting step in protein folding is the *cis-trans* isomerization of the X-Pro peptide bonds. Nearly all peptide bonds in proteins are in the *trans* configuration except for about 7% of the X-proline bonds (X can be any amino acid) which are in *cis* (*40*). To form the correct stereoisomer, the C-N bond of certain proline linkages must be rotated by 180°. This process is slow and often determines the overall folding kinetics as is the case with ribonuclease T1 (*25*). Enzymes with peptidyl-prolyl *cis-trans* isomerase activity have been shown to accelerate folding by catalyzing the rate limiting isomerization step. So far, two classes of peptidyl-prolyl *cis-trans* isomerases (PPI) have been distinguished: enzymes inhibited by cyclosporin (*25*) and those inhibited by the synthetic compound FK506 (*41, 42*). Both compounds are potent immunosupressants indicating a strong link between protein folding and the regulation of the immune response. Recently it was shown that at least one class of PPI's has been conserved from gram-negative bacteria to man suggesting that it must be playing an important role in cell physiology (*43*). However, preliminary studies have failed to demonstrate a relation between PPIase and the formation of inclusion bodies in *E.coli* (*10*).

Although post-translational modifications influence the rate of protein folding, the main driving forces that determine the three dimensional structure of proteins do not involve the formation of covalent bonds (more information on the forces that determine folding can be found in a recent review by Dill, *44*). Since the cell employs enzymatic mechanisms for the catalysis of reactions critical to polypeptide folding, it is reasonable to expect that some processes for facilitating non-covalent interactions must also be in place. Nevertheless, the idea that proteins interact and guide the proper folding of newly synthesized polypeptides is fairly recent and did not become accepted immediately. In 1983 Haas and Wabl demonstrated that the protein BiP is required for the assembly of immunoglobulins in the endoplasmic reticulum (*45*). Subsequent studies have shown that BiP belongs to a much more general class of folding-assisting proteins termed chaperonins (*46*). According to a general definition proposed by Ellis and Hemmingsen in 1989 (*47*), chaperonins are "proteins whose role is to mediate the folding of certain other polypeptides and, in some instances their assembly into oligomeric structures but which are not components of these final structures". Most, but not all, are induced by elevated temperatures and can be classified within three families of heat shock proteins (hsp's) based on molecular weight: hsp60 or the GroEL family, hsp70 and hsp90 (*1, 48,49*). Table 1 lists the known chaperonins and foldases of *Escherichia coli* and their functions. Chaperonins are involved in several important cellular functions including, the assembly of multimeric proteins, assembly of phages, protection of proteins from thermal deactivation (hence the induction of many chaperonins under temperature stress conditions), protein export from the cytoplasm and protein turnover. Furthermore, there is evidence that dnaK, an hsp70 member, can even mediate the ATP reactivation of aggregated proteins (*50,51*). The GroEL-GroES complex may also be able to exhibit a similar activity. The common theme in all these processes is a polypeptide folding/unfolding transition and this is where chaperonins come into play. The diverse functions of chaperonins are best illustrated into the following two examples.

The assembly of multimeric enzymes requires very specific orientation of two or more partially folded subunits. The encounter and precise interaction of two large and complex macromolecules in solution is normally a low probability event. As a result, *in-vitro* in the absence of additional factors, the correct assembly of protein subunits

Table 1. *Escherichia coli* Foldases and Chaperonins

Name	Molecular Weight, (d) Subunit/Oligomer	Functions
GroEL	58,000/812,0000	Forms complex with GroES Secretion; Phage assembly
GroES	10,000/70,000	Proteolysis; Reactivation of associated proteins (?)
DnaK	69,000	Secretion; Proteolysis; Deaggregation of aggregated proteins
SecB	17,000/119,000¶	Secretion
Trigger Factor	63,000	Secretion; Regulation of cell wall synthesis
PapD	- ¤	Assembly of pili
Peptidyl-prolyl *cis-trans* isomerase	18,500/ ?	Isomerization of Xaa-Pro peptide bonds

¶ Approximate M.W. based on gel filtration chromatography
¤ M.W. has not been reported

is inefficient and therefore non-specific aggregation of the polypeptide chains predominates, at least at physiological temperatures (*48,52*). This role of chaperonins was demonstrated in the assembly of the active 16mer of the chloroplast ribulose bisphosphate decarboxylase, or rubisco (Goloubinoff *et al.*, this volume, *53,54*). The formation of active rubisco in chloroplasts is facilitated by the rubisco binding protein. Similarly, the production of active rubisco in *E.coli*, requires the GroEL-GroES chaperonin complex. It turns out that the rubisco binding protein and the bacterial chaperonin are homologous. The assembly of rubisco by the GroEL and GroES chaperonin complex, requires ATP and K$^+$ ions. GroES associates into a heptameric complex whereas GroEL forms a 14mer consisting of two stacked rings of seven subunits each. Not surprisingly, the assembly of the GroEL complex itself is an autocatalytic process (*55*). Very little is known about how chaperonins recognize and bind a variety of structurally diverse substrates or how they facilitate the assembly process. Based on studies with BiP and hsc70, two folding catalysts from eucaryotic cells, Rothman and colleagues have proposed that the role of chaperonins is to retain the subunit polypeptides in a partially unfolded conformation necessary for assembly (*48*).

Chaperonins have a distinctly different role in protein secretion: they prevent the premature folding of the polypeptide chain in the cytoplasm and keep the protein in an "export competent conformation", *i.e.* sufficiently unfolded to be able to pass through the cytoplasmic membrane. For example, the GroEL chaperonin binds to and arrests the folding of the precursor form of secreted proteins, *i.e.*containing the signal sequence (*56*). Hydrolysis of ATP is required to release the precursor protein and allow it to continue folding. Interestingly, when GroEL was incubated with the folded precursor of ß-lactamase in the absence of ATP *in-vitro* it caused unfolding which could be reversed when ATP was added (*56*). A similar result has been observed with dihydrofolate reductase (Gatenby *et al.*, in press). In addition to GroEL, at least three other chaperonins bind to protein precursors during transport (Table 1). Although chaperonins have somewhat overlapping specificities for protein substrates, some are more important than others (*57*). In *Escherichia coli* GroEL is essential for cell viability whereas mutants deficient in SecB or dnaK have been isolated but grow considerably more slowly.

A relation between chaperonin-polypeptide interactions and the formation of inclusion bodies of recombinant proteins has been widely speculated (*1,10*) but no conclusive studies have been published so far. It is possible that aggregation occurs when a heterologous protein expressed in bacteria is either unable to bind to the host cell chaperonins or binds very tightly preventing dissociation and further interaction with newly synthesized molecules. Alternatively, even when proper binding can take place, the chaperonin concentration within the cell may be insufficient to cope with proteins expressed at high levels. Large differences in the extent of inclusion body formation are frequently observed when a protein is expressed in different *E. coli* strains with very similar genotypes and growth rates. The extent of aggregation in a particular strain may be related to the level of synthesis of chaperonin proteins. As more information on intracellular folding becomes available, it may be possible to manipulate the extent of inclusion body formation. For example, heterologous chaperonins could be cloned in *E. coli* to facilitate the folding of desired products or the synthesis of native chaperonins may be amplified to meet the increased demand when a foreign protein is overproduced.

The most extensively studied model for the study of folding and aggregation *in vivo* is the tailspike endorhamnosidase of phage P22 in S*almonella typhimurium* (Mitraki *et al.* and Chen and King, this volume). The ratio of aggregated to native tailspike increases at elevated temperatures and is further enhanced by certain amino acid substitutions. These observations indicate that the extent of protein aggregation in the cell can be controlled by growing the cells at suboptimal temperatures or by selecting for mutants that exhibit a decreased tendency to form inclusion bodies. The formation of inclusion bodies is often suppressed significantly by growing the cells at low temperatures (58,59). Furthermore it has been shown that the culture pH, the addition of ions and non-metabolizable sugars in the growth medium can influence the extent of soluble protein production (7). However, there are many proteins whose folding is not affected by the fermentation conditions. Recently, Wetzel and colleagues developed a very versatile selection scheme for the selection of interferon-γ mutants which exhibit increased solubility in the cell relative to the wild type protein. Some mutants exhibited antiviral activity comparable to the authentic interferon-γ but were produced in soluble form even when they were expressed at a high level (60, R. Wetzel, personal communication). The isolation of folding mutants with altered solubility characteristics is a promising method for preventing the formation of inclusion bodies.

Characteristics of Inclusion Bodies

Although there are numerous reports on the aggregation of different proteins in bacteria and yeasts, there is surprisingly little information about the composition and physicochemical properties of inclusion bodies. In *Escherichia coli* inclusion bodies can form in one of two locations, the cytoplasm or the periplasmic space (61). Bowden *et al.* (62) have shown that the characteristics of the protein aggregates in *E. coli* depend on how the protein is expressed. Constructs coding for ß-lactamase with the wild type or the OmpA signal sequence and with a complete deletion of the signal peptide were expressed at a high level causing the formation of inclusion bodies. The first two polypeptides accumulated in inclusion bodies in the periplasmic space and the latter in cytoplasmic aggregates. The inclusion bodies have distinctly different sizes and morphologies as evident from scanning electron microscopy observations (Figure 1). These differences extend to the surface morphology, composition and probably the conformation of the aggregating polypeptides.

Although inclusion bodies are generally amorphous, expression of insecticidal proteins in *E. coli* has been shown to result in the formation of crystalline, bipyramidal inclusion bodies in the cytoplasm when the cells were grown at 30°C (63). Growth at at 37°C induced the production of amorphous, spherical structures. The foreign protein corresponded to 70% of the total protein of these aggregates compared to 90% in the bipyramidal inclusion bodies formed at 30°C.

The size distribution and density of inclusion bodies have been studied only for two proteins, calf prochymosin and γ-interferon (64). The mean particle sizes of γ-interferon and prochymosin inclusion bodies were 0.81μm with a standard deviation of 0.17μm and 1.28μm with a standard deviation of 0.46μm, respectively. The buoyant density of the γ-interferon and prochymosin inclusion bodies was proportional to the density of the suspension medium indicating a relatively high voidage within the particles. The void fraction (=void volume/total volume) was 70% for γ-interferon and 85% for prochymosin.

The conformation of the polypeptide chains in the aggregates has not been investigated. In general, inclusion bodies are held together by non-covalent forces, mostly hydrophobic interactions. The cytoplasm of *E. coli* contains a rather high concentration of glutathione which normally prohibits the formation of disulfide bonds (40). As a result, proteins requiring disulfide bond formation for folding usually aggregate when they are expressed in the *E. coli* cytoplasm. Careful isolation of the inclusion bodies by sucrose gradient centrifugation showed that ß-lactamase derived polypeptides constitute more than 95% of proteinaceous material of the aggregates (62). These results indicate that in some cases other cellular proteins are not incorporated into the aggregate and therefore the formation of these structures must be an extremely specific process. Furthermore, chaperonins such as GroEL or SecB could not be detected in the inclusion bodies by immunoblotting indicating that they do not associate with aggregating polypeptides.

A

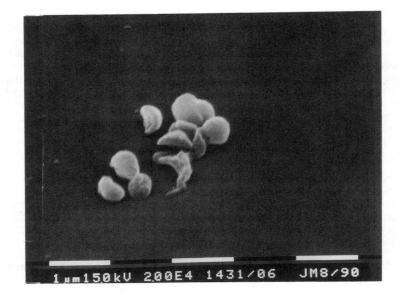

B

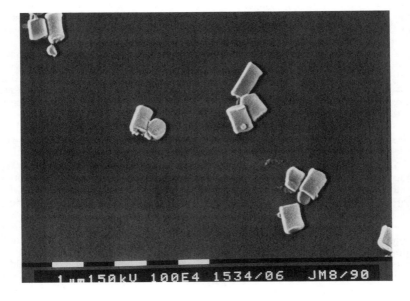

Figure 1. Scanning Electron Micrographs of β-lactamase inclusion bodies from *Escherichia coli*. Strain RB791 containing plasmid ptac11 which allows the overexpression of the wild type ß-lactamase or plasmid pGB1 which codes for the mature polypeptide without its signal sequence (62) were grown in minimal media. 24 hours after inoculation the cells were harvested by centrifugation, lysed and the inclusion bodies purified by sucrose gradient centrifugation. **A**: Inclusion bodies formed by the wild type ß-lactamase in the periplasmic space. **B**. Aggregates formed in the cytoplasm of RB791(pGB1) cells. Bar is equal to 1 μm.

Recovery of Proteins from Inclusion Bodies

There are many advantages to expressing recombinant proteins in inclusion body form from a downstream processing perspective. The protein of interest usually represents more than 50 % of the total protein contained in refractile bodies. The initial purification steps involve cell disruption and separation of the insoluble refractile material from soluble cellular components. Such processes are considered to be relatively simple. Once the inclusion bodies have been isolated, they are solubilized in the presence of strong denaturants and further purified in the denatured state. Purification sequences for inclusion body proteins generally require fewer steps than sequences for comparable proteins expressed in soluble form, which tends to save time and reduce losses.

The most important step of the inclusion body protein recovery scheme is the refolding of the denatured protein to form a biologically active product. Refolding can be relatively simple, for small monomeric proteins, or quite complicated when the protein consists of more than one polypeptide chains or contains several disulfide bonds. Difficult refolding processes can result in overall low recovery yields of active protein. This limits the use of inclusion body processes for the production of some recombinant products.

A general outline of an inclusion body protein recovery/purification scheme is presented in Figure 2. For a given protein, optimal conditions for each individual step have to be established and are a function of the composition of the starting material and the characteristics of the protein including protein size, the presence of inter and/or intramolecular disulfide bonds, the number and types of subunits in the protein molecule, and the presence of prosthetic groups.

The first step in the purification scheme is the release of inclusion bodies from the cells. Cells are concentrated by either centrifugation or crossflow filtration and are subsequently resuspended in a buffer of suitable ionic strength (~0.4 - 0.6 M salt). Cells are then lysed using mechanical techniques such as homogenization or by chemical or enzymatic methods such as treatment with lysozyme and EDTA. Soluble cellular materials are removed from the inclusion body preparation by cycles of centrifugation and resuspension in buffer (65). Soluble materials may also be removed by diafiltration (66), which decreases the fixed and operating costs and is more amenable to scale-up. At the completion of this step, the resulting preparation contains essentially all the inclusion body protein with a small amount of contaminating cell debris. Differential centrifugation in a sucrose gradient may be used to remove contaminants such as cell debris and membrane proteins. In cases where the desired protein is partially produced in soluble form, its recovery in the inclusion body fraction can be enhanced by treating the cells with non-polar solvents (such as phenol or toluene) or detergent prior to disruption. This cell killing step has been successfully used to increase the recovery of human growth hormone from recombinant *E. coli* inclusion bodies (67-69).

The purified inclusion body fraction is then solubilized and the residual insoluble material is pelleted and discarded. Solubilization of inclusion bodies is accomplished by using a combination of denaturants to disrupt non-covalent interactions and thiol reagents or sulfitolysis to break disulfide bonds.

The most commonly used denaturing agents are guanidine hydrochloride (GuHCl), urea, and ionic detergents. Chaotropic denaturants such as guanidine hydrochloride or urea, act by disrupting noncovalent bonds between polypeptide chains in the native molecule (70). As was mentioned above, unfolding is a cooperative process. There is some disagreement as to the precise mechanism of unfolding by chaotropic agents (20). Binding of the denaturant to protein molecules in the unfolded conformation is favored at high solute concentration (71). Typical

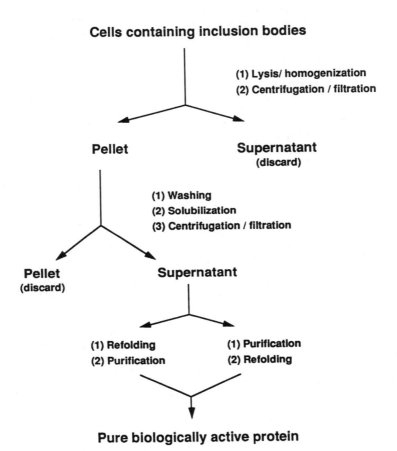

Figure 2. Inclusion body protein recovery scheme.

concentrations used to solubilize proteins are 4-7 M for guanidine hydrochloride and 5-10 M for urea.

Factors affecting protein denaturation in urea solutions differ in several respects from those influencing guanidine hydrochloride denaturation. Proteins in urea solutions are affected by the pH and ionic strength as a result of the electrostatic interactions between charged groups of the protein. These parameters are not as important in guanidine hydrochloride solutions which have a high ionic strength. Urea also has a tendency to undergo decomposition, especially at elevated temperatures (71). The reactive product of this reaction is cyanate ions which can react readily with free amines causing irreversible modification of the protein. Chemical modification of proteins in urea solutions is particularly important at neutral or alkaline pH.

While the mode of action of detergents and organic solutes is similar in some respects, the interaction between protein and reagent is significantly different for each type of denaturing agent. The binding between protein and detergent molecule is very strong and the conformational change between native and unfolded states occurs at low detergent concentrations. Detergents are used at concentrations in the range of 0.01 to 2 percent by weight. Detergents, particularly SDS, are generally used to solubilize inclusion bodies as a pretreatment of samples for gel electrophoresis. However, they are not used for the recovery of active inclusion body protein because their can bind to the protein irreversibly (72).

While denaturants are used to disrupt non-covalent interactions during solubilization, reducing conditions are utilized to disrupt disulfide bonds. Thiol reagents such as ß-mercaptoethanol (BME) and dithiothreitol (DTT) in the presence of a chelating agent such as EDTA are generally used as reducing agents in the cleavage of disulfide bonds. Typical concentrations are 0.1 M for DTT and 0.1-0.3 M for BME (73). An alternative method involves the use of sulfitolysis in the solubilization step, followed by sulfonate deletion in the refolding step (68). Sulfitolysis results in the transformation of the protein into its S-sulfonate form and is carried out by treating the protein with a mixture of a sulfite and a weak oxidizing agent.

When inclusion body protein accumulation levels are high (> 50% total cell protein), removal of cell debris and soluble cellular materials may not be necessary. Inclusion body protein may be directly solubilized by treating the cell concentrate with a strong denaturing solution, such as 4 M guanidine hydrochloride (74). This alternative strategy drastically reduces fixed and operating costs by eliminating the cell disruption and centrifugation operations. However, it requires an initial separation step to remove the contaminating materials prior to renaturation. Although in most inclusion body protein recovery processes solubilization and renaturation occur in separate steps, it is possible to solubilize, oxidize and refold a protein in a single step. The inclusion body protein is placed in a buffer of sufficient denaturing power to keep all the intermediate configurations during oxidation/refolding in solution. Urokinase (67) and methionyl bovine somatotropin (Storrs and Przybycien, this volume) have been recovered from inclusion body preparations by this procedure.

A different approach to a one-step inclusion body protein recovery process has been presented by Hoess et al. (12). Simultaneous solubilization and renaturation can be accomplished by incubating the inclusion body preparation with a strong ion exchange resin. This approach was demonstrated with the recovery of three proteins of different molecular size and complexity by incubating the inclusion bodies with a strong anionic resin (Q-Sepharose). The proteins recovered where the SV40 T antigen peptide Th, human interleukin 2 and the retroviral v-*myb* oncoprotein. It was also shown that the Th peptide can be effectively recovered from inclusion bodies by incubation with a strong cationic resin (S-Sepharose), indicating that adsorption of the protein to the ion exchanger is not essential for simultaneous solubilization and refolding. The mechanisms responsible for these phenomena are not yet known (12).

Refolding of inclusion body protein

In theory, renaturation may be accomplished by removal of the denaturing agent. However, in practice, the problem is far more complex and suboptimal renaturation can often lead to protein aggregation and/or inactivation with a low recovery of correctly folded protein.

The degree of protein aggregation is dependent on environmental parameters. Often aggregation is reduced when the pH of the medium is far removed from the isoelectric point of the protein. However, much more complex relations between the solution pH and the degree of protein aggregation have been reported (75). Aggregation generally increases with increasing temperature due in part to an increased probability of collision between protein molecules at elevated temperatures. This behavior has been observed with many proteins including ovalbumin (76), phenylalanyl - tRNA synthetase (77) and F_{ab} fragments (78).

Protein concentration probably has the most important effect on aggregation. Since it is an intermolecular process, it exhibits a reaction order equal to or greater than two (30). As a result, refolding processes are conducted at low protein concentrations, generally in the range of 10 - 100 μg/ml (73, 79). The specificity of the protein-protein contacts in aggregation is an open question. For a long time the association of polypeptides during refolding was considered a non-specific process. However, London et al. (80) observed that the aggregation of tryptophanase following refolding from a 3M guanidine-HCl solution is not affected by the addition of bovine serum albumin or crude extract from E. coli cultures. They concluded that foreign proteins do not interact with tryptophanase monomers during refolding and aggregation is caused by incorrect but specific binding of the polypeptide chains. The specificity of aggregation was also observed with phosphoglycerate kinase (81). Finally, Brems and coworkers demonstrated that the formation of protein aggregates in renaturation experiments arises from specific interactions among partially folded intermediates (82-84). They showed that a folding intermediate of bovine growth hormone forms rapidly upon refolding from urea solutions. This intermediate has an extended structure in which a hydrophobic segment located between the amino acids 110-127 is exposed to the solvent. At protein concentrations in excess of 0.2 mg/ml, intermolecular hydrophobic interactions lead to the formation of a dimeric intermediate. This dimeric species is the first step in the aggregation pathway (83). Protein association is enhanced when a lysine at position 112 is substituted with a hydrophobic leucine (84). On the other hand, the addition of a fragment of bovine growth hormone consisting of amino acids 96-133 to the refolding mixture prevents intermolecular interactions and the subsequent aggregation steps. Apparently, this peptide binds to the exposed hydrophobic domain of bovine growth hormone and protects it from associating with other partially folded protein molecules.

Although aggregation is the predominant means by which proteins become inactivated during refolding, several other inactivation pathways have also been observed. Proteins can be inactivated by thiol-disulfide exchange or alteration of the primary structure by chemical modification of amino acid side chains. In addition, refolded proteins may be inactive due to the absence of prosthetic groups and metals or because of improper association of the subunits in multimeric proteins (79).

Proteins which are difficult to fold often contain disulfide bonds. However, the presence of disulfide bonds does not necessarily constitute a difficult refolding process. Salmon growth hormone has been successfully refolded without the addition of thiol reagents or other additives which promote thiol-disulfide exchange (85). Less than 2% of bovine growth hormone was found to be in an aggregated form when refolding was done without the addition of reagents which promote thiol-disulfide exchange (86). Of course this result is certainly not representative of all proteins which contain disulfide bonds.

The formation of cysteine pairs can occur through oxidation by air, through the presence of both reducing reagents and oxidized thiol compounds, and through the activity of proteins that catalyze the interchange of disulfide bonds. Air oxidation is the oldest technique used to renature thiol-containing proteins. Air oxidation of reduced proteins can be catalyzed by trace metal ions. Both Cu^{2+} and Co^{2+} have been observed to be effective catalysts for the oxidation of cysteines in a number of proteins. The rate and yield of correctly folded protein following air oxidation is usually disappointing. For example the renaturation of reduced RNAase takes several hours and results in only 10% yield of active protein (87).

The rearrangement of disulfide bonds during protein refolding can be accelerated by facilitating thiol-disulfide exchange. Both the addition of a small quantity of reducing or oxidizing agent or the addition of a mixture of reduced and oxidized thiols act in a similar manner. The basic thiol exchange reaction involves the ionized form of the thiol and is therefore pH dependent. Typical reagents used in these reactions include ß-mercaptoethanol (BME), dithiothreitol (DTT), and glutathione; the latter is the most commonly used. Typical concentrations are 1-5 mM thiol and 0.01-0.5 mM disulphide reagent. Optimum oxidation usually occurs with a 10-fold molar excess of the reduced thiol reagent relative to the molar concentration of cysteines in the protein (73). Disulfide-sulfhydryl interchange can also be accelerated by utilizing the reshuffling enzyme protein disulfide isomerase (88). Thioredoxin has also been reported to facilitate disulfide isomerization (87). However, in our experience, thioredoxin does not accelerate the refolding of RNAase with respect to air oxidation but not relative to redox buffers (Cronau and Georgiou, unpublished observations).

In addition to pH, temperature and protein concentration, other factors that have to be considered in maximizing refolding yields include protein purity and the rate of solubilizing agent removal. Although it has been suggested that the removal of contaminants prior to refolding may result in an increase in the overall renaturation yield (89), Buchner and Rudolph (78) found that the homogeneity of the inclusion body preparation has no effect on the recovery of functional recombinant F_{ab} fragments. Nevertheless, many processes involve an initial purification step before removal of the solubilizing agent (69,90). When guanidine hydrochloride and detergents are used for solubilization, the initial purification step is gel filtration chromatography. Ion exchange chromatography can also be used when urea is the solubilizing agent, resulting in the simultaneous removal of nucleic acids, phospholipids and electronegative contaminant proteins (91). Hydrophobic and reverse phase chromatography has also been used for the purification of proteins in the denatured state (92).

In order to maximize refolding yields, conditions must be established so that the kinetic competition between folding and inactivation/aggregation reactions favors folding. Since the kinetic constants in the refolding pathway are a strong function of the concentration of solubilizing agents, the rate at which these agents are removed affects the rate of accumulation of folding intermediates, and thus, protein inactivation and aggregation. The selection of an optimal rate of solubilizing agent removal and its implications are discussed by Vicik and De Bernardez-Clark (this volume).

The removal of the solubilizing agents can be accomplished by a variety of techniques. These include dilution, dialysis, gel filtration, diafiltration and immobilization on a solid support. Dilution is the simplest way to initiate refolding and has been used extensively (93,94). However, downstream processing volumes are large, increasing purification costs. Dialysis allows for separation of substances by means of a semipermeable membrane. In principle, smaller molecules and ions diffuse through the membrane readily, whereas larger molecules and colloidal particles pass through slowly or not at all. Examples of proteins which have been refolded using dialysis are the *E. coli* tryptophanase (80), bovine growth hormone, urokinase and FMD virus protein type A (67-69, 95,96). While the refolding of polypeptide chains using dialysis can result in biologically active proteins, this diffusion limited process is very slow. Because of the characteristically slow processing times, dialysis is not used in large scale production processes.

The refolding of a given protein through a reduction in denaturant concentration may also be accomplished through gel filtration. The continuous addition of buffer during size exclusion chromatography acts to dilute the denaturant. The protein exiting a column is normally more concentrated and purer than the material loaded. However, it should be noted that the loss of protein solubility or aggregate formation during denaturant removal could cause flow restrictions within the column. Amons and Schrier (97) have effectively used gel filtration to remove SDS from proteins after the protein had been dissociated from the detergent with propionic acid.

Diafiltration is traditionally used for buffer exchange, the new buffer is continuously added while the old one is removed at an equivalent rate, thereby keeping the protein in a constant volume environment. Diafiltration through a semipermeable membrane selectively removes denaturant and other small molecules while retaining the protein of interest. These attributes make it attractive for large scale processing. Diafiltration has been used in the refolding of prorennin (68) and interferon-ß (98).

In order to reduce interactions between protein molecules, and thus prevent the formation of aggregates, proteins can be refolded while immobilized on a solid support. Sinha and Light (99) have refolded trypsinogen with 60-70% yields and trypsin with 50% yield by first binding the protein to agarose beads and subsequently refolding it in the presence of mercaptoethanol or a mixture of reduced and oxidized glutathione. Creighton (100) proposed the use of ion-exchange resins as solid supports for the refolding of proteins denatured by urea. Folding was induced by gradually altering the solvent. High yields were obtained in the refolding of cytochrome c but it was observed that binding of the protein to the resin interferes with the refolding of BPTI. Furthermore, only 10% renaturation yield was obtained in the refolding of α-lactalbumin and lysozyme. This was attributed to a combination of protein aggregation and interference in folding by adsorption of the protein to the resin.

Another technique to reduce the aggregation of proteins during refolding is through treatment with citraconic anhydride. Addition of this reagent changes the positively charged amino group of residues such as lysine to a negatively charged carboxylate ion. This change reduces protein-protein interactions which depend on electrostatic effects. The citraconic anhydride-protein interaction is readily reduced at low pH (101). Light (102) has used this approach to refold trypsinogen.

Refolding yields can be improved by utilizing "defensive reversible denaturation". This technique has been shown to be effective in restoring the activity of denatured interferon preparations (103). It was observed that interferons denatured by a variety of techniques can be partially refolded by incubation in low concentrations of urea. When SDS was added to the urea-interferon solutions prior to renaturation, full activity of the interferons could be restored. Stewart et al. (103) interpreted the stabilizing effect of SDS as "defensively" denaturing the protein solution.

When protein aggregation occurs under native conditions, refolding yields can be improved by incubation of the protein in non-denaturing concentrations of chaotropic agents. Refolding yields of chymotrypsinogen from 6M GuHCl were optimized by incubating the protein solution in 1.2M GuHCl in the presence of reduced and oxidized glutathione (104). Similarly, Tsuji et al. (105) found that oxidation of interleukin 2 during refolding has to be carried out at intermediate concentrations of GuHCl (2 M) in order to minimize aggregation. Incubation at intermediate denaturant concentrations during oxidation was also used to improve TPA refolding yields (106).

Labilizing and stabilizing agents have also been used to enhance renaturation yields. Preferential destabilization of incorrectly folded species can be accomplished by addition of labilizing agents, such as L-arginine, during refolding. Buchner and Rudolph (78) studied the effect of labilizing agents on the refolding of recombinant fragments. They found that the most pronounced effect on yield was achieved by adding L-arginine to the renaturation buffer. Concentrations of L-arginine in the range of 0.35 - 0.5 M resulted in a 60% increase in the refolding yield. Conversely, intermediates can be protected from self-association by stabilizing agents. Cleland and Wang (*107*) showed that bovine carbonic anhydrase B refolding rates can be significantly increased by adding polyethylene glycol (PEG) to the refolding buffer. The presence of PEG also prevented the formation of aggregates during refolding. Hershenson *et al.* (*98*) obtained high renaturation yields of interferon-ß when stabilizing agents were added to the refolding buffer. Among the stabilizing agents utilized are detergents (such as Trycol, Durfax, Pluraface, and Tween) and sugars, alcohols and polyols including glycerol, sucrose, isopropanol and PEG of 100-2000 MW. Finally, Valax and Georgiou (this volume) showed that the addition of sucrose or other sugars in the refolding buffer increases the recovery of active ß-lactamase following dialysis from guanidine-HCl solutions.

Conclusions

The formation of protein aggregates or inclusion bodies is a widespread phenomenon that occurs not only during the expression of recombinant proteins but also under physiologically relevant conditions. It is generally thought that protein aggregation *in-vivo*, as well as *in-vitro*, results from the failure of a polypeptide intermediate to complete folding leading to self-association. The accumulation of proteins in inclusion bodies is related to the polypeptide sequence, expression rate, availability of chaperonins and the growth environment. To an extent all these factors may be manipulated to influence the solubility of the protein within the cell.

Although the formation of inclusion bodies may seem undesirable, there are many advantages to expressing recombinant proteins in refractile form, especially in downstream processing. However, the application of inclusion body processes has been limited by the low yields obtained in the refolding step. Since the factors that determine protein folding are not completely known, no general rules or procedures have been established for recovery of inclusion body proteins in high yields. Numerous techniques have been explored to improve refolding yields and some guidelines for developing an efficient refolding procedure have been established based on these techniques. As our understanding of protein folding improves, the recovery of functional proteins from the aggregated state is likely to become increasingly simplified.

References

1. Horwich, A.L.; Neupert, W.; Hartl F.-U. *Trends in Biotechnol.* **1990**, *8*, 126-131.
2. Takahashi, N.; Hayano, T.; Suzuki, M. *Nature* **1988**, *337*, 473-475.
3. Williams, D. C.; van Frank, R.M.; Muth, W.L.; Burnett, J.P. *Science* **1982**, *215*, 687-689.
4. Marston, F.A.O. *Biochem. J.* **1986**, *240*, 1-12.
5. Leemans, R.; Remaut, E.; Fiers, W. *J. Bacteriol.* **1987**, *169*, 1899-1904.
6. Schein, C. H.; Kashiwagi, K.; Fujisawa, A.; Weissman, C. *Bio/Technology* **1986**, *4*, 719-725.
7. Bowden, G.; Georgiou, G. In *Recombinant DNA Technology and Applications*; Prokop, A.; Bajpai, K.R.; Ho, C. Eds.; **1990**, McGraw-Hill, pp. 333-356.
8. Berndt, N.: and Cohen, P.T.W. *Eur. J. Biochem.* **1990**, *190*, 291-291.
9. Prouty, W. F.; Karnovsky, M.J.; Goldberg, A.L. *J. Biol. Chem.* **1975**, *250*, 1112-1122.
10. Schein, C. H. *Bio/Technology* **1989**,*7*, 1141-1149.
11. Sharma, S. K. *Separ. Sci. and Technol.* **1986**, *21*, 701-726.

12. Hoess, A.; Arthur, A.K.; Wanner, G.; Fanning, E. *Bio/Technol.* **1988**, *6*, 214-1217.
13. Tokatlidis, K.; Dhurjati, P.; Millet, J.; Beguin, P.; Tremel, R.; Longin, R.; Aubert, J.-P. *Paper presented at the 199th ACS National Meeting*, Boston, MA **1990**.
14. Shibui, T. *et al. J. Biotechnol.* **1990**, *17*, 109-120.
15. Baneyx F.; Georgiou, G. In *Pharmaceutical Biotechnology v.2*, Manning, M.; Ahern, T. Eds.; Plenum Press, in press.
16. Cheng, T.E.; Kwoh, T.J.; Soltvedt, B.C.; Zipser, D. *Gene* **1981**, *14*, 121-130.
17. Babbitt, P.C.; West, B.L.; Buechter, D.D.; Kuntz, I.D.; Kenyon, G.L. *Bio/Technology* **1990** 8, 945-949.
18. Wittrup, K.D.; Mann, M.B.; Fenton, D.M.; Tsai, L.B.; Bailey, J.E. *Bio/Technology*, **1988**, *6*, 423-426.
19. Rudolph, R.; Fischer, S.; Mattes, R. *International Patent Application WO 87/02673* **1987**.
20. Creighton, T.E. *Proteins* **1983**, Freedman & Co.
21. Baldwin, R. L. *Trends in Biochem. Sci.* **1989**, *14*, 291-294.
22. Miranker, A.; Radford, S.E.; Karplus, M.; Dobson, C.M. *Nature*, **1991**, *349*, 633-636.
23. Staley, J.P.; Kim, P.S. *Nature* **1990**, *344*, 685-688.
24. Kim, P.S.; Baldwin, R.L. *Ann. Rev. Biochem.* **1990**, *59*, 631-660.
25. Fischer, G.; Schmid, F.X. *Biochemistry* **1990**, *29*, 2205-2212.
26. Udgonkar, J.B.; Baldwin, R.L., *Nature* **1988**, *335*, 694-699.
27. Roder, H.; Elöve, G.A.; Englander, S.W. *Nature*, **1988**, *335*, 700-704.
28. Bycroft, M.; Matouschek, A.; Kellis, J.T. Jr.; Serrano, L.; Fersht, A. *Nature*, **1990**, *346*, 488-491.
29. Robson, B.; Pain, R. H. *Biochem. J.* **1976**, *155*, 325-330.
30. Zettlmeissl, G.; Rudolph, R.; Jaenicke, R. *Biochemistry* **1979**, *18*, 5567-5571.
31. Tsou, C.L., *Biochemistry*, **1988**, *27*, 1809-1817.
32. Luger, K.; Hommel, U.; Herold, M.; Hofsteenge, J.; Kirschner, K. *Science* **1989**, *243*, 296-210.
33. Liu, G.; Topping, T.B.; Randall, L.L. *Proc. Natl. Acad. Sci. U.S.A.* **1989**, *86*, 9213-9217
34. Hardy, S.J.; Randall, L.L. *Science* **1991**, *251*, 439-443.
35. Laminet, A.A.; Plückthun, A. *EMBO J.* **1989**, *8*, 1469-1477.
36. Bowden, G.A.; Georgiou, G. *J. Biol. Chem.* **1990**, *265*, 16760-16766.
37. Schein, C.H. *Bio/Technology* **1990**, *8*, 308-317.
38. Freedman, R.B. *Cell* **1989**, *57*, 1069-1072.
39. Gilbert, H.F. *Adv. Enzymol.* **1990**, *63*, 69-172.
40. Stewart, D.E.; Sarkar, A.; Wampler, J.E. *J. Mol. Biol.* **1990**, *214*, 253-260.
41. Topschung, M.; Wachter, E.; Mayer, S.; Schönbrunner, E.R.; Schmid, F.X. *Nature* **1990**, *346*, 674-677.
42. Harrison, R.K.; Stein, R.L. *Biochemistry* **1990**, *29*, 3813-3817.
43. Liu, J.; Walsh, C.T. *Proc. Natl. Acad. Sc. U.S.A.* **1990**, *87*, 4028-4032.
44. Dill, K.A.; *Biochemistry* **1990**, *29*, 7133-7155.
45. Haas, I.G.; Wabl, M. *Nature* **1983**, *306*, 330-334.
46. Hemmingsen, S. M.; Woolford, C.; van der Vies, S.M.; Tilly, K.; Dennis, C.T.; Georgopoulos, C.; Hendrix, R.W.; Ellis, R.J. *Nature* **1988**,*333*, 330-334.
47. Elis, R.J.; Hemmingsen, S.M. *Trends Biochem. Sci.* **1989**, *14*, 339-342.
48. Rothman, J. E. *Cell* **1989**, 59, 591-601.
49. Schlesinger, M.J. *J. Biol. Chem.* **1990**, *265*, 1211-1214.

50. Gaitanaris, G.A.; Papavasiliou, A.G.; Rubock, P.; Silverstein, S.J.; Gottesman, M.E. *Cell* **1990**, *61*, 1013-1020.
51. Skowra, D.; Georgopoulos, C.; Zylicz, M. *Cell* **1990**, *62*, 939-942.
52. Viitanen, P.V.; Lubben, T.H.; Reed, J.; Goloubinoff, P.; O'Keefe, D.P.; Lorimer, G.H. *Biochemistry* **1990**, *29*, 5665-5671.
53. Goloubinoff, P.; Gatenby, A.A.; Lorimer, G.H. *Nature* **1989**, *337*, 44-47
54. Goloubinoff, P.; Christeller, J.T.; Gatenby, A.A.; Lorimer, G.H. *Nature* **1989**, *342*, 884-889.
55. Lissin, N.M.; Venyaminoy, S.Y.; Girshovich, A.S. *Nature* **1990**, *348*, 339-342.
56. Laminet, A.A.; Ziegelhoffer, T.; Georgopoulos, C.; Plückthun, A. *EMBO J.* **1990**, *9*, 2315-2319.
57. Lecker, S.; Lill, R.; Ziegelhoffer, T.; Georgopoulos, C.; Bassford Jr., P.J.; Kumamoto, C.A.; Wickner,W. *EMBO J.* **1989**,*8*, 2703-2709.
58. Schein, C. H.; Noteborn M.H.M. *Bio/Technology* **1988**, *6*, 291-294
59. Kopetzki, E.; Schumacher, G.; Buckel, P. *Mol. Gen. Genet.* **1989**,*216*, 149-155.
60. Wetzel, R.; Perry, L.J.; Veileux, C.; Chang, G. *Protein Eng.* **1990**, *3*, 611-623.
61. Georgiou, G.; Telford, J.N.; Shuler, M.L.; Wilson, D.B. *Appl. and Environ. Microbiol.* **1986**, *52*, 1157-1161.
62. Bowden, G.A., Paredes, A.M.; Georgiou, G. *Bio/Technology*, submitted.
63. Oeda, K.; Inouye, K.; Ibuchi, Y.; Oshie, K.; Shimizu, M.; Nakamura, K.; Nishioka, R.; Takada, Y.; Ohkawa, K. *J. Bacteriol.* **1989**,*171*, 3568-3571.
64. Taylor, G.; Hoare, M.; Gray, D.R.; Marston, F.A.O. *Bio/Technology* **1986**, *4*, 553-557.
65. Lowe, P.A.; Rhind, S.K.; Sugrue, R.; Martson, F.A.O. In *Protein Purification: Micro to Macro* **1987**, Alan R. Liss, Inc., 429-442.
66. Forman, S.; Swartz, R.; De Bernardez, E.; Feldberg, R. *J. Membrane Sci.* **1990**, *48*, 263-279.
67. Olson, K.C. *U.S. Patent 4,518,526* **1985**.
68. Wetzel, R.B. *U.S. Patent 4,599,197* **1986**.
69. Builder, S.M.; Ogez, J.R. *U.S. Patent 4,620,948* **1986**.
70. Tanford, C. *Adv. Protein Chem.* **1968**, *23*, 122-282.
71. Privalov, P.L. *Adv. Protein Chem.* **1979**, 33, 167-241.
72. Ghelis, C.; Yon, J. In *Protein Folding*, Jaenicke, R.; Ed. **1982**, Academic Press, pp. 220-373.
73. Jaenicke, R.; Rudolph, R. In *Protein Structure. A Practical Approach*, Creighton, T.E.; Ed. **1990**, IRL Press, 191-222.
74. Kung, H.-F. *US Patent 4,476,049* **1984**.
75. Mulkerrin, M.G.; Wetzel, R. *Biochemistry* **1989**,*28*, 6556-6561.
76. Holme, J. *J. Phys. Chem.* **1963**, *67*, 782-788.
77. Goerlich, O.; Holler, E. *Biochemistry* **1984**,*23*, 182-190.
78. Buchner, J.; Rudolph, R. *Bio/Technology* **1991**, *9*, 157-162.
79. Mozhaev, V.V.; Martinek, K. *Enzyme Microb. Technol.* **1982**,*4*, 299-309.
80. London, J.; Skrzynia, C.; Goldberg, M.E. *Eur. J. Biochem.* **1974**, *47*, 409-15.
81. Mitraki, A.; Betton J.-M.; Desmadril, M.; Yon, J. *Eur. J. Biochem.* **1987**, *163*, 29-34.
82. Havel, H. A., E. W. Kauffman, S. M. Plaisted, Brems, D.N. *Biochemistry* **1986**, *25*, 6533-6538.
83. Brems, D. N. *Biochemistry* **1988**, *27*, 4541-4546.
84. Brems, D. N.; Plaisted, S.M.; Havel, H.A.; Tomich, C.S.C. *Proc. Natl. Acad. Sci. USA* **1988**, *85*, 3367-3371.

85. Sekine, S.; Mizukami, T.; Nishi, T.; Kuwana, Y.; Saito, A.; Sato, M.; Itoh, S.;
     Kawauchi, H. *Proc. Natl. Acad. Sci. USA* **1985**, *82*, 4306-4310.
86. Gill, J.A.; Sumpter, J.P.; Donaldson, E.M.; Dye, H.M.; Souza, L.; Berg, T.;
     Wypch, J.; Langley, K. *Bio/Technology* **1985**, *3*, 643-646.
87. Pigiet, V.P.; Schuster, B.J. *Proc. Natl. Acad. Sci. USA* **1986**, *83*, 7643-
     7647.
88. Tang, J.G.; Wang, C.C.; Tsou, C.L. *Biochem. J.* **1988**, *255*, 451-455.
89. Shire, S.J.; Bock, L.; Ogez, J.; Builder, S.; Kleid, D.; Moore, D.M.
     Biochemistry **1984**, *23*, 6474-6480.
90. Marston, F. A. O. In *DNA Cloning: A Practical Approach, Vol. III*, Glover,
     D.M.; Ed. **1987**, IRL Press, Oxford, pp. 59-89.
91. Thatcher, D.R. *Biochem. Soc. Trans.* **1990**, *18*, 234-235.
92. Knuth, M.W.; Burgess, R.R. In *Protein Purification: Micro to Macro* **1987**,
     Alan R. Liss, New York, pp. 279-305.
93. Hager, D.A.; Burgess, R.R. *Anal. Biochem.* **1980**, *109*, 76-86.
94. Martson, F.A.O.; Lowe, P.A.; Doel, M.T.; Schoemaker, J.M.; White, S.;
     Angal, S. *Bio/Technol.* **1984**, *2*, 800-804.
95. Langley, K.E.; Berg, T.F.; Strickland, T.W.; Fenton, D.M.; Boone, T.C.;
     Wypych, J. *Eur. J. Biochem.* **1987**, *163*, 313-321.
96. Winkler, M.E.; Blaber, M.; Bennet, G.I.; Holmes, W.; Vehar, G.A.
     Bio/Technol. **1985**, *3*, 990-1000.
97. Amons, R.; Schrier, P.I. *Anal. Biochem.* **1981**, *116*, 439-443.
98. Hershenson, S.; Shaked, Z.; Thomson, J. *U.S. Patent 4,961,969* **1990**.
99. Sinha, N.K.; Light, A. *J. Biol. Chem.* **1975**, *250*, 8624-8629.
100. Creighton, T.E. *In Protein Structure, Folding, and Design* **1986**, Alan R.
     Liss, Inc., 249-257.
101. Dixon, H.B.F.; Perham, R.N. *Biochem. J.* **1968**, *109*, 312-314.
102. Light, A. *Biotechniques* **1985**, *3*, 298-305.
103. Stewart, W.E.; De Somer, P.; De Clercq, E. *Prep. Biochem.* **1974**, *23*, 383-
     393.
104. Orsini, G.; Goldberg, M.E. *J. Biol. Chem.* **1978**, *253*, 3453-3458.
105. Tsuji, T.; Nakagawa, R.; Sugimoto, N.; Fukuhara, K. *Biochem.* **1987**, *26*,
     3129-3134.
106. Sarmientos, P.; Duchesne, M.; Defenle, P.; Boizau, J.; Fromage, N.; Delporte,
     N.; Parker, F.; Lelievre, Y.; Mayaux, J.-F.; Cartwright, T. *Bio/Technol.*
     1989, 7, 495-501.
107. Cleland, J. and Wang, D.I.C. *Bio/Technol.* **1990**, 8, 1274-1278.

RECEIVED March 22, 1991

Chapter 2

Physical Methods and Models for the Study of Protein Aggregation

Catherine H. Schein

Laboratory for Organic Chemistry, Swiss Federal Institute of Technology (E.T.H.), CH8092 Zürich, Switzerland

Of course the chicken had to come first,
for how else would the egg know what it had to look like?

Two of the major problems in refolding proteins from "inclusion bodies", the insoluble particles formed in bacterial cells producing recombinant proteins, are precipitation during the refolding process and chemical alteration of the protein during the fermentation, denaturation step, or purification. Precipitation can occur at any point in protein purification, so it is important to define the solubility limits of the protein as soon as practicable during the isolation. Physical methods for following the course of aggregation and for detection of modified forms of recombinant proteins are summarized.

As the limited solubility of recombinant proteins is a major problem in their purification and use[1], we would like to determine why some proteins form aggregates rather than their native soluble state. Aggregation can occur at any point in the purification and storage of a protein, but is frequently ignored unless a precipitate is seen. Besides interfering with many purification steps, aggregation of proteins lowers process yields and the specific activity (units/mg pure protein) attainable. Many "new" spectral methods can be used to determine the presence of aggregation before visible precipitation occurs. Spectral measurements can also be used to discriminate between native and denatured forms of the protein and thus to follow the course of refolding or to check for structural lability during storage.

Table I summarizes physical methods available for following the course of aggregation, determining the structure of protein precipitates, and characterizing the native state structures of proteins. Most of these techniques are old workhorses that have been taken from chemistry and physics and adapted for biochemistry. The solution spectra of proteins are very complex due to the large number of different side chains and the size of the molecules themselves. Fortunately, a few researchers find such systems a challenge; protein NMR would not be possible today without the theoretical underpinnings developed by physicists/physical chemists and the biochemists [2,3] who put the methods to the test. Recent improvements in vibrational spectroscopy, mass spectrometry, and laser light scattering at both the measurement and the interpretation level mean that they can provide much better characterization of protein structure in solution or solids. To take advantage of these methods, lab equipment in the 1990's will need to be updated and biotechnologists must become a bit more conversant in the relevant theory.

0097–6156/91/0470–0021$06.00/0

Table I. A comparison of physical methods for monitoring changes in protein structure and interprotein interactions involved in aggregation with respect to the protein requirements, impedence of measurement by common buffers, and the type of information obtainable about the protein in solution or precipitate

Method	[Protein] required*	Sample preparation/ buffer interference**	Gives information for a protein on: precipitate/complex mol.wt./shape	Secondary structure	side chain interactions
UV-absorption	low	Only for protein in solution	no	no	(Aromatics)
Fluorescence	low	"	no	no	(Aromatics)
Circular dichroism	moderate		no	yes	no
Electron microscopy	low	Fixation/staining required	yes	no	no
Electron diffraction	low to high	2-D crystals or membrane patches	yes	yes	possibly
Fourier Transform- Infrared Absorption (FTIR)	high	Can also be used for precipitates	no	yes	possibly
Laser light scattering:					
Elastic (classical, static)	moderate	Scattering methods are usually not greatly affected by the buffer.	yes	no	not specific
Quasi-elastic (dynamic)	moderate		yes	no	not specific
Raman	high	Sample fluoresence may interfere	no	yes	possibly
Raman resonance	moderate	with Raman resonance measurement of precipitates.	no	yes	possibly
Neutron scattering	high	Beam time is limited	yes	some	no
Mass spectrometry (MS)	varies	Usually protein is bound to a solid phase (like nitrocellulose)	Very accurate m.w.	possibly	possibly
Nuclear magnetic resonance (NMR)	high	Most solvents interfere with measurements.	possibly	yes	yes
X-ray diffraction	high	Crystals required	yes	yes	yes

* Protein concentration required for measurement. Low is in the nM-μM range, high means 0.1mM or higher. Note that highly concentrated protein solutions are usually required to obtain crystals and that a lot of protein is wasted during the process of crystallization. Thus one needs to start with 100 mg or so of purified protein for X-ray analysis, although only a few μgs may find their way to the beam.
**Buffers used for sample preparation for all methods should be sterile and ultrafiltered (0.45μ or smaller pore diameter filter) to remove particulate matter. Buffer requirements for various methods are discussed in the references given.

This does not mean that physical measurements can stand alone. The characteristics of a protein that may lead to aggregation are precisely those that are hardest to determine with spectroscopic methods: patches of amino acids that may attract other areas of the protein; conformation of the main chain that limit flexibility; acceptance of alternate hydrogen bond donors from the solvent in place of the intraprotein contacts of the native state; groups on the surface that form covalent or ionic linkages if brought within a close enough distance to another molecule; hydrophobic or ionic attachment of side chain residues to the glass or plastic receptacle[4]; proteolysis of N- and C- termini due to traces of protease in the preparation; deamination or oxidation of side chains. For this reason, physical methods give the most meaningful information when combined with information from site directed mutagenesis or antibody studies.

Comparative sequence analysis of even well described inter-protein interactions (i.e., from X-ray structure of multi-domain or subunit proteins) has only recently been possible; initial results have however shown some common features of quaternary interactions[5]. There is hope that interactions in naturally self-associating proteins will be similar to those that lead to irreversible interactions in refolded proteins, and that one can then use this knowledge to prevent precipitation. For example, fibrin clots can be dissolved by adding the tetra-peptide Gly-Pro-Arg-Pro, the amino acids at the N-termini of fibrinogen molecules after thrombin cleavage[6]. A related peptide, Arg-Gly-Asp-Ser, from the carboxy- terminus of fibrinogen, inhibits platelet aggregation[7].

Model systems for the study of irreversible protein aggregation

There are many ways that a protein can form contacts with itself; thus results from one model system may be very difficult to transfer to another. For example, an early assumption that inclusion bodies of recombinant proteins were due to hydrophobic interactions similar to those of membrane protein aggregates led to the use of purification protocols based on those used to dissolve proteins in cell membranes. But one generally cannot use membrane protein purification protocols for the isolation of IB protein, for the simple reason that the aggregates differ in their structure and the mechanism of formation[8]. Native membrane proteins have hydrophobic patches on their surface that enable them to interact with lipids in the membrane; well chosen detergents replace these interactions and allow the protein to be isolated[9]. Detergents encourage proteins to unfold by serving as an alternate binding site for hydrophobic residues that should be "buried" within the native protein; thus they can prevent the proper refolding of proteins after denaturation.

A single mechanism will not describe even the "controlled" aggregation of proteins in nature. A single viral coat protein, for example, may form several different contacts with itself and other proteins, depending on its final position within the shell structure[10]. Indeed, the original postulate of "quasi-equivalent" binding at lattice points in virus capsules was modified to "non-equivalency" when the first structure was solved at atomic dimensions. For example, the coat of Tomato bushy stunt virus consists of 180 identical subunits of a 43 kD protein which self-interact at lattice points in at least three distinct ways[11].

Although cell membrane proteins interact with each other and lipid components to allow maintenance of a negatively charged, hydrophobic bilayer that permits the entry of nutrients but keeps the nucleic acids of the cell from escaping[12], the contacts formed are not exclusively "hydrophobic". Several parts of the sequence must interact with a series of other molecules in order for the protein to be properly placed within the membrane: amino terminal leader peptides target the protein to the membrane, "stop-transfer" and "insertion domains" (hydrophobic stretches separated by polar regions) allow interaction with lipids, and other amino acid sites (many as yet unidentified) serve as sites for attachment of linkers molecules like fatty acids[13].

In view of the diversity of possible inter-protein contacts, the best method for dealing with aggregation problems is to study the protein at hand. The most important step is to prepare a good phase diagram of the protein in terms of its solubility as a function of salt, pH and other co-solvents [8]. When this is done, laser light scattering techniques can then be used to follow aggregation or vibrational spectroscopic methods used to characterize the spectrum of the protein in different environments. The following examples suggest ways of combining methods to obtain maximum information.

Physical methods for studying aggregation

Traditional:
Of the commonly used spectral methods, UV and fluorescence spectroscopy of non-modified proteins are limited as the spectra usually consist of fairly wide bands, attributable largely to the tryptophan content of the protein. Visible spectra can be used to follow an increase in turbiditry with time but give little qualitative data. Stopped-flow turbidimetry has been used to derive a kinetic model for the precipitation of α-chymotrypsin by different salts[14]. *Circular dichroism (CD)* is the most commonly used method to determine peptide secondary structure, but can only determine what percent of the structure of a larger protein is in a particular conformation. Even this determination is open to debate, as there is no clear "random coil" reference structure for use in deconvoluting the spectrum[15]. Vibrational CD spectroscopy in the amide 1 band region may give more information[16]. CD is most useful for following the course of denaturation under conditions where aggregation can be ignored or for establishing that a protein has refolded properly when a reference spectrum of the native protein is available. For example, CD has been used to demonstrate that addition of polyols increases the concentration of guanidinium hydrochloride needed to denature lysozyme or ribonuclease[17], and (coupled with UV difference spectra) to show conformational changes in ovalbumin during freeze denaturation[18]. Residual structure in fragments of *Staphylococcal* nuclease was demonstrated by comparing their CD spectra with the spectrum of the full length wild type protein[15].

The traditional method to determine the presence of aggregation after differential centrifugation is *gel electrophoresis* (also coupled with immunoblotting or isoelectric focusing[19]); inclusion bodies have been shown to contain non-SDS/mercaptoethanol separable protein-polymers[20,21] by this method. The low pH and Ca^{2+} dependent aggregation of secretogranin II was observed *in vitro* by centrifuging samples after dialysis against various buffer solutions and observing the protein content of the pellet and supernatant fractions[22]. (The authors concluded that the observed aggregation may be essential for protein sorting to secretory granules, as the *in vivo* packaging of secretogranin could be prevented by neutralizing the pH of acidic compartments of the cell with ammonium chloride.) Methods based on gel electrophoresis are convenient to use for initial solubility determination as they require very little protein and can be used with complex protein samples.

Another method which requires little protein is *electron microscopy (EM)*, which has long been the tool of choice for the characterization of cell organelles and non-crystalline complexes.   The method was used to demonstrate *in vitro* reconstruction of organelles such as *Salmonella* flagellar filaments (phosphotungstic acid stained preparations)[23] and microtubules (frozen-hydrated or glutaraldehyde/tannic acid fixed and stained)[24]. In the latter study, the authors were able to distinguish three classes of microtubules differing by only one protofilament, and both preparation methods showed the same results.   Assuming that the precipitate is stable to the sample preparation conditions, stained thin sections, scanning, or cryo-electron micrographs can be used to characterize its size and shape. A three dimensional interpretation of two dimensional data can be obtained from micrographs made before and after tilting the specimen (which can be fixed on a special tiltable platform). For example, a structure at 37 Å resolution for ribosomes from rabbit reticulocytes (which have not yet been successfully crystallized like the ribosomes of thermophilic and halophilic bacteria discussed later) was calculated from electron micrographs of uranyl-acetate stained single-particle specimens[25].

One recent study followed the growth of lysozyme crystals by stereo-electron microscopy . The crystals were frozen at various stages of growth (the growth rate of the crystals was regulated by varying the concentration of protein in the mother liquid) in liquid xenon and platinum-carbon replicas were prepared of the growing crystal surface. Nucleation sites were then calculated (using a simple model) from the superposition pattern of the monomolecular-layers observed. The nucleation sites per crystal so estimated were exponentially related to the lysozyme concentration in the mother liquor. The same technique could also be used to follow the aggregation of tomato bushy stunt virus. [26]

Exciting results have recently been obtained using *electron cryo-microscopy* to generate *electron diffraction* patterns for obtaining medium- to high-resolution molecular images of crystalline specimens. For example, purple membranes from bacteria are a natural two-dimensional crystalline array of lipids and bacteriorhodopsin. As the array is only one unit cell thick, it is not suitable for X-ray analysis but yields electron diffraction patterns with spots to beyond 4Å resolution. By tilting the sample, one alters the focus to obtain a series of patterns that represent progressive planes at various heights of the specimen.   From these overlaying planes, a 3-dimensional model, primarily based on the bulkier and aromatic amino acid side chains, has been suggested for the structure of bacteriorhodopsin in the membranes[27]. A projection map to 4Å of an α-helical coiled-coil protein from insect eggs has also been determined from high-resolution electron microscopy of microcrystals[28].

"New" methods:

Because even a small protein will have many different absorbing (or resonating) parts, *IR (and Raman)* spectral data represent an average of signals coming from diverse sites in the protein. The overlap of these signals can make the spectrum so broad as to appear meaningless.   The spectra can, however, be deconvoluted, by fitting a linear superposition of model spectra to the averaged spectrum to extract the overlapping waveforms. Deconvolution requires knowing the frequencies where the individual parts of a protein will absorb. This determination has been attempted using model proteins where the structure has been determined by X-ray crystallography. Deconvoluted spectra in the amide I band region (between 1620 and 1700 cm^{-1}), have been used to make secondary structure interpretations in terms of % α-helix, ß-sheet, etc. [29,30]. It has been suggested that the spectral interpretation can be made more accurate by factor analysis[31]. Deconvolution of other spectral areas may yield more information on side chain interactions as well, but the model systems must be developed. Site specific isotopic labelling of enzymes like that used for NMR[32,33] can in theory be used to resolve selected peaks in the spectrum; this approach has already been applied to [13]C- and [2]H-labeled phospholipids[34].

Highly concentrated protein solutions, thin protein films, or precipitate suspensions in various solvents can be used for *Fourier transform infrared spectroscopy (FTIR)* . Great care should be taken in interpreting deconvoluted spectra; the basic data is still only one curve. Using *IR difference spectra* avoids the possible errors of deconvolution, but the protein environment during the measurement must be absolutely controlled to obtain meaningful data.

"Resolution enhancement techniques" may give IR an advantage over CD in distinguishing fine differences in structure. For example, the X-ray structure of the Ca^{2+} binding protein α-lactalbumin contains both relatively long stretches of 3_{10}-helix and normal α-helices. FTIR spectra of α-lactalbumin in D_2O show a band at 1639 cm^{-1} that is probably attributable to the 3_{10}-helices as well as bands at 1651 and 1659 cm^{-1} that are attributable to α-helices[35]. In a study of the pressure-induced denaturation and "formation of a white gel" of chymotrysinogen, changes in the amide II band at 1550 cm^{-1} (which includes the C-N stretching and N-H in-plane bending vibrations) during hydrogen exchange time course were followed. The amide II band switches to 1457 cm^{-1} when the amide groups are deuterated; if the protein is placed in deuterated water (D_2O), the peak position shifts as the protein denatures and more of the internal ("buried") hydrogen atoms exchange. If the pressure was changed gradually, the protein denatured at a higher pressure (5.5 kbar) than after rapid compression increase (ca. 3.7 kbar). The authors concluded that the secondary structure alterations induced by pressure denaturation are different than those observed after thermal denaturation of both chymotrypsinogen and lysozyme. [36]. One drawback of the method is the very high protein concentration used (100 mg/ml), which is difficult to attain for most proteins and would certainly give aggregation problems if thermal denaturation studies were attempted.

Oriented membrane proteins, membrane fragments, or proteins in a micelle environment can also be analyzed using *"infrared attenuated total reflection spectroscopy"* (IR-ATR). Here, the sample is coated onto an optically transparent germanium plate instead of being sandwiched between cells. The infrared beam is reflected within the plate and up through the sample. Multiple internal reflections along the plate greatly increases the sensitivity of the method. As ATR measurements are less impeded by water than transmission IR, the changes in the protein with environmental alterations are easily followed by means of difference spectra[37,38].

Laser light scattering methods are all based on the same principle: when a beam of light impinges on a molecule, a small fraction of the light is scattered by the molecules. Light scattered at approximately the incoming frequency is measured in classical ("elastic") and quasi-elastic (dynamic) light scattering. An even smaller fraction of the light is scattered at frequencies higher (anti-Stokes) and lower (Stokes) than the incident beam. These scattered intensities are measured in *Raman scattering*, which has many variations. Obviously, the sensitivity of scattering measurements is dependent on the quality of the incident beam. Lasers give a coherent, high intensity, monchromatic beam; one limit on the method is that the intensity should not be high enough to damage the sensitive biological specimen.

For *elastic (Classical, static) light scattering*, the scattering of a beam of polarized laser light as it passes through the sample is measured and the obtained "scatter factor" (scattered intensity or Raleigh ratio) can be used to calculate both the weight average molecular weight of the particles in solution as well as the radius of gyration. To obtain the molecular weight of a particle, one measures the scattering factor at several different scattering angles (θ) and then plots the scattering factor as a function of the θ. Extrapolating back to $\theta=0$ will yield the molecular weight.

In a study of monoclonal antibody aggregation in the presence of antigen, the estimates of the molecular weight and the radius of gyration in solution, as measured by classical light scattering, agreed very well with those estimated from electron micrographs of the same aggregates[39]. The aggregation of low density lipoprotein (LDL) induced by acetylation, carbamylation, maleylation, or oxidation, was evaluated by laser light transmission fluctuation. The authors report that aggregated LDL, but not unmodified monomer, stimulates the uptake of cholesteryl esters in arteriosclerotic cells [40].

The difference between elastic and "quasielastic" measurements is that in the latter, small changes in the frequency due to the translational ("Brownian") movement of the scattering particles are also measured. The broadness of the intensity distribution of the emitted light for frequencies around the primary monochromatic beam frequency is directly related to the diffusion coefficient of the particles, which can then be related to the hydrodynamic radius if a model for the particle shape is available[41],[42]. Dynamic light scattering can thus be used to follow the kinetics of particle coagulation by following the decrease in diffusion coefficient as the particle size increases[43].

Dynamic light scattering data can be used to determine particle diffusion coefficients as a function of time, but there are problems with determining the shape. Related information from other methods, eg. electron microscopy or classical light scattering, that suggest a particular geometry for the aggregates and a theoretical model, is needed in order to use the diffusion coefficient to determine the dimensions of the particles. For example, the aggregation of bovine α-crystallin, purified from fetal tissues, has been studied with both classical and dynamic light scattering. For data interpretation, the particle geometry was assumed to be a wormlike chain, with a chain diameter based on EM, and a value for the persistence length which was treated as a free parameter. A quantitative comparison of the light scattering data and the data from EM indicated that the sample preparation methods used for (negative staining) EM had induced some particle breakdown, thus illustrating the advantage of using several characterization methods (P. Schurtenberger and R.C. Augusteyn, manuscript in preparation). Alternatively, if one only wants to compare scalar measurements for a series of samples, dynamic light scattering measurements of antibody-antibody complex formation have been interpreted by treating the particles as having "fractal" dimensions rather than a defined geometric shape[44].

Dynamic (quasi-elastic) laser light scattering is one of the few methods that can be used to study the course of aggregation in any solvent. This method is particularly suited to study samples in different buffers as long as one can measure the viscosity accurately. Fibrin is a good model system as there are several ways to induce the build up and break down of aggregates and the site of aggregation has been biochemically characterized. A combination of dynamic and static light scattering measurements done after induction of fibrin clotting by addition of thrombin or reptilase under controlled conditions yielded data that could be used to model fibrin polymerization. The results were correlated with antibody studies of samples taken at various points during the measurements that followed the conversion of fibrinogen to fibrin monomers[6].

A major question in protein chemistry is: why does one salt (or salt concentration) induce crystallization while most other conditions lead to precipitation of a concentrated protein solution? A quasi-elastic light scattering study of the effect of salts on lysozyme and concanavalin A precipitation concluded that salts that encourage precipitation lead to an increase in macromolecule polydispersity and size with increasing concentration. Salts that lead to crystallization, on the other hand, do not encourage an increase in particle size and the translational diffusion coefficients do not decrease as the salt concentration is raised[45]. Light scattering studies of solutions of canavalin in a fixed salt concentration (where the protein was known to form rhombohedral crystals) showed that at pH 7, three clearly defined domains of behavior were seen as a function of protein concentration. At low concentration (0.12% w/v), the particle diameter measured was 14-36 Å and incubated samples remained clear for 104 days. In zone B, (>0.12 to ≤0.2% w/v), particle size and polydispersity increased very rapidly with small increases in protein concentration. The average particle diameter was 181 Å but the range was from 44-214Å, and incubated samples contained precipitate and "deformed aggregates" . Above this protein concentration, the particle sizes were between 200 and 290 Å, with no further defined increase in size with concentration. At 0.3% w/v protein, the mean particle size was 238 Å, which approximates the diameter of an aggregate of 8 canavalin trimers when placed at each lattice point of a rhombohedral cell. As solutions with >0.3% protein formed crystals at these conditions, the authors assume such aggregates are the nucleation sites for crystallization. With non-optimal NaCl concentrations, the aggregate size increased much more slowly as a function of protein concentration[46].

Raman Spectroscopy has been used extensively to study the quaternary interactions of hemoglobin and to compare the secondary structures of proteins under various conditions[47,48]. It has the disadvantage that protein solutions must be highly concentrated to obtain meaningful data, as the scattered light intensity correlates directly with the protein content of the sample. For example, 10-20% solutions of α-chymotrypsin and chymotrypsinogen were used for a Raman spectroscopic analysis of secondary changes induced by pH and pressure[49]. However, precipitate samples can be analyzed directly. Raman spectroscopy has been used to show a correlation between the salt type and concentration used for precipitation of α-chymotrypsin and the disordering of the secondary structure of the protein caused by the salts[50]. When the method was extended to another 11 proteins, the general result was that the precipitated protein contained a higher percentage of ß-sheet structure as determined by the amide I band intensity and location; the effect was more pronounced when the chaotrophic salt KSCN was used compared to the "structure stabilizing" salt, Na_2SO_4[51].

Further details on the shape and flexibility of aggregates can be obtained from *neutron diffraction* studies, but beam time is not easily obtained. *Small angle neutron scattering (SANS)* was originally suggested as a way to determine the amount of water bound by a protein in different solvents, but the data error is large when the particle radius is <100 Å . SANS has been used in combination with ultracentrifugation to determine the particle weight of malate dehydrogenase from a halophilic bacteria in different salt solutions. The data suggest that in 1-2 M NaCl solution, the protein is a stable dimer and binds much more water and NaCl than in low salt solution, where the "salt-loving" protein unfolds [52]. Neutron scattering can be used in combination with light scattering or electron microscopy to study whole biological aggregates. For example, scattering from suspensions of casein micelles (made from fresh bovine milk and suspended in differing concentrations of ^2H_2O) showed that there was an inflection in the intensity vs scattering factor curve, which, taken together with EM data, was interpreted to mean that the structure was one of closely packed micelle monomer subunits[53]. RecA protein complexes with itself and with DNA under various conditions have also been studied with EM and neutron scattering; as the authors work in Grenoble they could do both techniques simultaneously, which made comparisons easier[54]. Neutron diffraction of the crystals of ribosome subunits that are discussed later may be used as a direct method of phasing (Eva Pebay-Peyroula and Michel Roth, Institute Laue-Langevin, Grenoble, unpublished data).

Assuming one can crystalize the complex, very detailed structural information can be obtained by *X-ray analysis*. Several protein:protein complexes have been solved to high resolution as for example a complex of thermitase (279 amino acids) with the inhibitor Eglin C (70 amino acids) has been resolved to < 2 Å resolution[55] and a complex of human thrombin (two chains, 39 and 259 amino acids) and its inhibitor hirudin (65 amino acids) to 2.3 Å[56]. The recent 2.8 Å resolution crystal structure of an antibody with its peptide antigen basically confirmed the results of immunological mapping experiments[57]. Crystals of a complex of bacterially produced anti-lysozyme Fv region and lysozyme refract to 2.5 Å and should thus also yield high resolution structure suitable for analyzing the binding site[58]. The advantage of X-ray crystallography is that one obtains exact information about the residues in contact; the disadvantage is that one cannot be sure that the crystallization conditions have not significantly altered the structure of the complex. Thus even this method requires confirming data from other measurements.

The crystallization of large complexes is a formidable undertaking[59]. For example, the photosynthetic reaction center (crystals diffract to ~3Å)[60] and the nucleosome core particle (<5Å resolution with defined sequence DNA[61]) projects required many years of work on improving the crystallization conditions before any structural analysis could begin.   Crystals of profilin:actin were first reported in 1976; data from crystals diffracting to 1.8 Å resolution were published in 1989 (ATP hydrolysis seemed to be required for the efficient formation of crystals, and very careful transfer to a 3.2 M ammonium sulfate/ATP bathing solution was required to keep the crystals stable) [62]. Ribosomes from thermophilic microbes can be crystallized; most of the early crystals diffracted to >20 Å and were thus most useful for comparing the exterior structure with EM photos[63]. However, recent improvements have led to crystals that diffract to <5 Å which could be used, in combination with image reconstruction, to determine structural details of the ribosome[64].

On the technical side, synchrotron X-ray radiation is necessary for most protein complexes as they have large unit cells and weak diffraction patterns. It may be necessary to protect the crystals from the beam by freezing them; this was particularly valuable for obtaining data from ribosome crystals[65].

In the absence of crystals, *small-angle X-ray scattering* may be used to follow the course of precipitation. The method is valuable for studying very fast aggregation steps; a recent paper followed the rapid assembly of Tobacco mosaic virus protein induced by a temperature jump. The minimum counting time was 7.5 seconds[66]. Small angle X-ray and neutron scattering have been used to analyze the geometry of calmodulin in complexes with two peptides[67]. However, synchrotron radiation is only available in a few cities in the world; beam time is a precious commodity.

The other high resolution method for determining non-covalent bonding in proteins is *Nuclear magnetic resonance (NMR)*.   Even if one has access to a 500 or 600 mega-Hertz instrument, to obtain detailed information, either the size of the protein must be limited or the molecules must be site-specifically labeled. The advent of 3-D NMR[68,69] and even "4D" NMR for double labeled proteins [70] should allow the study of much larger complexes.   One method that has already yielded results for complexes uses 2-D NMR to observe the binding of (labeled) small molecules to an unlabeled protein.   Two examples are the binding of cyclophilin to cyclosporin[71] and the binding of an antibody fragment (Fab) to a peptide[72]. As the on/off rate of the peptide is short compared to the spin-lattice relaxation time of the Fab and peptide protons, a NOESY spectrum in the presence of a vast excess of peptide has extra cross peaks ("transfer NOE") which are due to magnetic exchange between the bound and free peptide fractions and are not present when the peptide and Fab are present in a 1:1 ratio [73]. The transfer NOE can thus be used to determine the peptide residues in contact with the protein.

One can follow protein aggregation even with low magnetic field NMR via the changes in water proton relaxation. Upon aggregation, water proton relaxation times decrease. These effects were originally attributed to release of "bound" water, but a recent study suggests that in reality the effective proton relaxation times of the protein itself change due to inefficient averaging during rotational motion, and that no information is given on the hydration state[74].

One can also obtain NMR spectra for proteins in micelles, which may allow the study of membrane protein structure in an environment approximating their native one. A combination of labels ($^{15}N,^{13}C$) was used for NMR studies of detergent-solubilized M13 coat protein. Although most of the resonances in the spectrum have not been assigned, there was clear indication that many of the protein residues had two distinct resonances of equal intensity. This was interpreted to mean (in combination with the results of sedimentation equilibrium, Raman and CD studies) that the protein was present in two conformers that represent the non-equivalent monomers of an asymmetric dimer[75]. NMR has also been used to determine the spatial structures of gramicidin A and ^{19}F-labeled bacteriorhodopsin fragments in a membrane-like milieu[76].

Methods for comparing recombinant proteins with their "native" counterparts

One must establish that refolded proteins have a conformation that is similar to the native structure. The most common method for doing this is to compare CD spectra (eg.,ref.[77]), as spectropolarimeters are widely available. However, other methods may yield more information. For example, Raman spectra, using the 500-550 cm^{-1} v_{S-S} region, can demonstrate the correct formation of di-sulfide bonds in the refolded protein[78]. NMR can also be used, as was recently demonstrated for hen egg white lysozyme (HEWL) produced in *Aspergillus niger*. The protein was purified from the culture supernatant; while the final preparation had virtually identical spectra compared to HEWL from egg white, the partially purified protein showed missing or irregular spots in "fingerprints" from COSY and NOESY spectra that disappeared after dialysis at pH 2. The authors concluded that while the purified protein did indeed have a conformation identical to the native protein, during production the protein bound to medium components[79].

Proteins for pharmaceutical use should not have modified side chains. The common method for detection of modified residues is to see a blank space instead of a peak in the HPLC elution trace of the amino acid analysis or a band at another position than the amino acid indicated by the gene sequence. This analysis will probably not detect modifications that affect only a small fraction of the protein. Elucidation of the identity of the modified side chains, if one can locate them, requires some astute chemistry. For example, in phycocyanine, N-methyl-asparagine ran exactly at the position of serine! Identification of the modified residue relied on detection of methyl amine released by acid hydrolysis of the protein[80].

New methods use combined HPLC/Mass spectrometry to identify modified amino acids. Purified recombinant human insulin-like growth factor separated into two peaks on reverse phase HPLC (C_{18} column/acidified water) even though other methods indicated it was completely pure. *Plasma desorption mass spectrometry*[81] of the individual peptides detected a single methionine sulfoxide molecule that was sufficient to decrease the hydrophobicity of the whole protein significantly[82]. Most of the oxidation occurred when the secreted fusion protein was cleaved with hydroxylamine under not strictly anaerobic conditions, but about 5% occurred during the *E. coli* fermentation.

Even proteolysis may be difficult to detect with the usual method, gel electrophoresis.   Clipping of the C-terminal arginine in erythropoietin was demonstrated by *FAB-MS (fast atom bombardment mass spectrometry)* of the C-terminal peptide[83]. Indeed, mass spectrometry will probably be one of the more important "new" tools in biochemistry of the '90s as methods for analysis of even fentomole quantities of proteins as large as 100 kD have been developed[84,85].

Conclusions

Protein precipitation during purification is just one aspect of a larger and more interesting question: how and why do proteins interact ? Accurate determination of secondary, tertiary, and quaternary protein structure requires combining information from various physical and biochemical methods. There are many physical methods now available for the determination of the structure of proteins and protein aggregates. Many of the methods discussed in this review are just being developed for use with biological materials. Although better spectrometers are being built all the time, the real problem is interpreting the data obtainable with the existing instrumentation. This will require better protein model systems and mathematical analysis of spectral data.   Most of all, improvements will come from biotechnologists actually using the methods.

Acknowledgements: I wish to thank the following E.T.H. experts: Peter Schurtenberger (Polymer Chemistry), laser light scattering; Gerhard Frank (Molecular biology), amino acid analysis of proteins; Walter Amrhein (Organic Chemistry), Mass spectrometry; Urs Fringeli (Physical Chemistry), IR of membranes, Werner Braun (Biophysics), NMR and mathematical methods in general; and Roger Brunne (Informatikgestützte Chemie) and my colleagues Kai Johnsson, Andy Tauer, and Monika Haugg for reading the manuscript.

References

[1]Schein, C.H. *Bio/Technology* **1989**, *7*, 1141-1148.

[2]Ernst, R.R., Bodenhausen, G., and Wokaun, A. *Principles of nuclear magnetic resonance in one and two dimensions*, Clarendon: Oxford, 1987.

[3] Wagner,G., Braun, W., Havel, T.F., Schaumann, T., Go, N., and Wüthrich, K. *J. Mol. Biol.* **1987**, *196*, 611-639   .

[4]Shirahama, H., Lyklema, J., and Norde, W. *J. Colloid Interf. Sci.*, **1990**, *139*, 177-187.

[5]Argos, P. *Protein Engineering* **1988**, *2*, 101-113.

[6]Dietler, G. Fibrin polymerization: a combination of light scattering with measurements of fibrinopeptide release. Diss. ETH No. 7819, 1985 (Department of Physics, Swiss Federal Institute of Technology, CH8092 Zürich).

[7]Krishnamurthi, S., Dickens, T.A., Patel, Y., Wheeler-Jones, C.P.D., and Kakkar, V.V. *Biochem. Biophys. Res. Comm.* **1989**, *163*, pp. 1256-1264.

[8]Schein, C.H. *Bio/Technology* **1990**, *8*, 308-316.

[9]Michel, H. *Trends Biol. Sci.* **1983**, *6*, 56-58.

[10]*Virus structure and assembly*; Casjens, S., Ed. Jones and Bartlett Publishers: Boston, MA 1985.

[11]Olson, A.J., Bricogne, G. and Harrison, S.C. *J. Mol.Biol.* **1983**, *171*, 61-93.

[12]Westheimer, F.H. *Science* **1987**,*235*, 1173-78.

[13]Dalbey, R.E. *Trends Bio. Sci.* **1990**, *15*, 253-257.

[14]Przybycien, T.M. and Bailey, J.E. *AIChE J.* **1989**, *35*, 1779-1790.

[15]Shortle, D. and Meeker, A.K. *Biochemistry* **1989**, *28*, 936-944.

[16]Drake, A.F., Siligardi, G., and Gibbons, W.A. *Biophys. Chem.* **1988**, *31*, 143-146.

[17]Gekko, K. and Ito, H. *J. Biochem.* **1990**, *107*, 572-577.

[18]Koseki, T., Kitabatake, N., and Doi, E. *J. Biochem.* **1990**, *107*, 389-394.

[19]van den Oetelaar, P.J.M., de Man, B.M., and Hoenders, H.J. *Biochim. Biophys. Acta* **1989**, *995*, 82-90.

[20]Schein,C.H. and Noteborn, M.H.M. *Bio/Technology* **1988**, *6*, 291-294.

[21]Albiges-Rizo, C. and Chroboczek, J. *J. Mol. Biol.* **1990**, *212*, 247-252.

[22]Gerdes, H. H., Rosa, P., Phillips, E., Baeuerle, P.A., Frank, R., Argos, P., and Huttner, W.B. 1989. *J. Biol. Chem.* **1989**, *264*, 12009-12015.

[23]Ikeda, T., Asakura, S., and Kamiya, R. *J. Mol. Biol.* **1989**, *209*, 109-114.

[24]Wade, R.H., Chrétien, D. and Job, D. *J. Mol. Biol.* **1990**, *212*, 775-786.

[25]Vershoor, A. and Frank, J. *J. Mol. Biol.* **1990**, *214*, 737-749.

[26]Durbin, S.D. and Feher, G. *J. Mol. Biol.* **1990**, *212*, 763-774.

[27]Henderson, R., Baldwin, J.M., Ceska, T.A., Zemlin, F., Beckmann, E., and Downing, K.H. *J. Mol. Biol.* **1990**, *213*, 899-929.

[28]Bullough, P.A. and Tulloch, P.A. *J. Mol. Biol.* **1990**, *215*, 161-173.

[29]Casal, H.L., Köhler, U. and Mantsch, H.H. *Biochim. Biophys. Acta*, **1988**, *957*, 11-20.

[30]Hester, R.E.. and Austin, J.C. in *Spectroscopy of Biological Molecules: New Advances*, Schmid, E.D., Schneider, F.W., and Siebert, F., Eds. John Wiley and Sons, Ltd.: Chichester, England, 1988, 3-10.

[31] Lee, D.C., Haris, P.I., Chapman, D. and Mitchell, R.C. *Biochemistry*, **1990**, *29*, 9185-9193.

[32]Fesik, S.W. *Nature* **1988**,*332*, 865-866.

[33]McIntosh, L. P. and Dahlquist, F.W. *Quat. Rev. Biophys.*, **1990**, *23*, 1-38.

[34]Blume, A. and Hübner, W. in *Spectroscopy of Biological Molecules: New Advances*, Schmid, E.D., Schneider, F.W., and Siebert, F., Eds. John Wiley and Sons, Ltd.: Chichester, England, 1988, 157-164. *

[35]Prestrelski, S.J., Byler, D.M., and Thompson, M.P. *Biochem.*, **1990**, in press.

[36]Wong, P.T.T. and Heremans, K. *Biochim. Biophys. Acta* **1988**, *956*, 1-9.

[37]Fringeli, U.P. in *Biologically Active Molecules*, Schlunegger, U.P., Ed. Springer-Verlag: Heidelberg, 1989, 241-252.

[38]Fringeli, U.P., Apell, H.J., Fringeli, M., and Läuger, P. *Biochim. Biophys. Acta*, **1989**, *984*, 301-312.

[39]Murphy, R.M., Slayter, H., Schurtenberger, P., Chamberlin, R.A., Colton, C.K. and Yarmush, M.L. *Biophys. J.* , **1988**, *54*, 45-56.

[40]Tertov, V.V., Sobenin, I.A., Gabbasov, Z.A., Popov, E.G., and Orekhov, A.N. *Biochem. Biophys. Res. Comm.* **1989**, *163*, 489-494.

[41]Galla, H.-J. *Spektroskopische Methoden in der Biochemie.* Georg Thieme Verlag: Stuttgart and New York, 1988.

[42]Berne, B.J. and Pecora, R. *Dynamic Light Scattering.* John Wiley and Sons: New York, 1976.

[43]Versmold, H. and Härtl, W. *J. Chem. Phys.*, **1983**, *79*, 4006-4009.

[44]Rarity, J.G., Seabrook, R.N., and Carr, R.J.G. *Proc. R. Soc. London A.* **1989**, *423*, 89-102. The authors define an object as having fractal dimensions when, if material is present at one point, the probability of finding material at some distance r in three dimensional space from that point is related to r^{d-3}, where d, the correlation fractal dimension, is not an integer.

[45]Mikol, V., Hirsch, E. and Giegé, R. *J. Mol. Biol.* **1990**, *213*, 187-195.

[46]Kadima, W., McPherson, A., Dunn, M.F., and Jurnak, F.A. *Biophys. J.* **1990**, *57*, 125-132.

[47]Williams, R.W. *J.Mol. Biol.*, **1983**, *166*, 581-603. ; also, Williams, R.W. *Meth. Enzym.* **1986**, *130*, 311-331.

[48]Rousseau, D.L. and Ondrias, M.R. In *Optical Techniques in Biological Research*; Rousseau, D.L., Ed.; Academic Press: Orlando, FL, 1984, pp. 65-132.

[49]Heremans, L. and Heremans, K. *Biochim. Biophys. Acta* **1989**, *999*, 192-197.

[50]Przybycien, T.M. and Bailey, J.E. *Biochim. Biophys. Acta* **1989**, *995*, 231-245.

[51]Przybycien, T.M. and Bailey, J.E. Secondary Structure perturbations in salt-induced protein precipitates. submitted to *Biochim. Biophys. Acta*, **1990**.

[52]Zaccai, G. and Eisenberg, H. *Trends Biol. Sci.* **1990**, *15*, 333-337.

[53]Stothart, P. H. *J. Mol. Biol.* **1989**, *208*, 635-638.

[54]DiCapua, E., Schnarr, M., Ruigrok, R.W.H., Lindner, P., and Timmins, P.A. *J. Mol. Biol.* **1990**, *214*, 557-570.

[55]Gros, P., Betzel, C., Dauter, Z., Wilson, K.S., and Hol, W.G.J. *J. Mol. Biol.* **1989**, *210*, 347-367.

[56] Rydel, T.J., Ravichandran, K.G., Tulinsky, A., Bode, W., Huber, R., Roitsch, C., and Fenton, J.W. *Science* **1990**, *249*, 277-280.

[57]Stanfield, R.L., Fieser, T.M., Lerner, R.A., and Wilson, I.A. *Science* **1990**, *248*, 712-719.

[58]Boulot, G., Eiselé, J-L., Bentley, G.A., Bhat, T.N., Ward, E.S., Winter, G., and Poljak, R.J. *J. Mol. Biol.* **1990**, *213*, 617-619.

[59]Huber, R. *EMBO J.* **1989**, *8*, 2125-2147.

[60]Deisenhofer, J. and Michel, H. *EMBO J.* **1989**, *8*, 2149-2170.

[61]Richmond, T.J., Searles, M.A., and Simpson, R.T. *J. Mol. Biol.* **1988**, *199*, 161-170.

[62]Schutt, C.E., Lindberg, U., Myslik, J. and Strauss, N. *J. Mol. Biol.* **1989**, *209*, 735-746.

[63]Trakhanov, S., Yusupov, M., Shirokov, V., Garber, M., Mitschler, A., Ruff, M., Thierry, J.C., and Moras, D. *J. Mol. Biol.* **1989**, *209*, 327-328.

[64]Yonath, A. and Wittmann, H.G. *Trends Biol. Sci.* **1989**, *14*, 329-335.

[65]Hope, H., Frolow, F., von Böhlen, K., Makowski, I., Kratky, C., Halfon, Y., Danz, H., Webster, P., Bartels, K.S., Wittmann, H.G., and Yonath, A. *Acta Cryst.* **1989**, *B45*, 190-199.

[66]Hiragi, Y., Inoue, H., Sano, Y., Kajiwara, K., Ueki, T., and Nakatani, H. *J. Mol. Biol.* **1990**, *213*, 495-502.

[67]Trewhalla, J., Blumenthal, D.K., Rokop, S.E., and Seeger, P.A. *Biochem.* **1990**, *29*, 9316-9324.

[68]Griesinger, C., Sørensen, O.W., and Ernst, R.R. *J. Magn. Reson.* **1987**, *73*, 574-579.

[69]Nagayama, K., Yamazaki, T., Yoshida, M., Kanaya, S., and Nakamura, H. *J. Biochem.* **1990**, *108*, 149-152.

[70]Kay, L.E., Clore, M., Bax, A., and Gronenborn, A.M. *Science* **1990**, *249*, 411-414.

[71]Wider, G., Weber, C., Traber, R., Widmer, H. and Wüthrich, K. Use of a double-half-filter in two dimensional ^{1}H NMR studies of receptor bound cyclosporin. *JACS* **1990**, in press.

[72]Levy, R., Assulin, O., Scherf, T., Levitt, M. and Anglister, J. *Biochem.* **1989**, *28*, 7168-7175.

[73]Anglister, J.   *Quaterly Reviews of Biophysics* **1990**, *23*, 175-203.

[74]Hills,B.P., Takacs, S.F., and Belton, P.S.   *Molecular Physics* **1989**, *67*, 919-937.

[75]Henry, G.D. and Sykes, B.D.   *J. Mol. Biol.* **1990**, *212*, 11-14.

[76]Bystrov, V.F., Arseniev, A.S., Barsukov, I.L., Lomize, A.L., Abdulaeva, G.V., Sobol, A.G., Maslennikov, I.V., and Golovanov, A.P. in *Protein Structure and Engineering*, Jardetsky, O., ed., Plenum Press: New York, 1989, 111-138.

[77]Kubota, N., Orita, T., Hattori, K., Oh-eda, M., Ochi, N. and Yamazaki, T. *J. Biochem.* **1990**, *107*, 486-492.

[78]Carey, P.R.   in *Spectroscopy of Biological Molecules: New Advances*, Schmid, E.D., Schneider, F.W., and Siebert, F., Eds. John Wiley and Sons, Ltd.: Chichester, England, 1988, 11-15.

[79]Archer, D.B., Jeenes, D.J., Mackenzie, D.A., Brightwell, G., Lambert, N., Redford, S.E., and Dobson, C.M. *Bio/Technology* **1990**, *8*, 741-745.

[80]Rümbeli, R., Suter, F., Wirth, M., Sidler, W., and Zuber, H. *Biol. Chem. Hoppe-Seyler* **1987**, *368*, 1401-1406.

[81]Roepstorff, P.   *Acc. Chem. Res.* **1989**, *22*, 421-427.

[82]Hartmanis, M.G.N. and Engstrom, Å. in *Techniques in Protein Chemistry*, Hugli, T.E:, Ed., Academic Press:San Diego, CA. **1989**, 327-333.

[83]Recny, M.A., Scoble, H.A., and Kim, Y.   *J. Biol. Chem.* **1987**, *262*, 17156-17163.

[84]Loo, J.A., Edmonds, C.G., Smith, R.D., Lacey, M.P., and Keough, T.   *Biomed. Env. Mass Spectrom.* **1990**, *19*, 286-294.

[85]Smith, R.D., Loo, J.A., Edmonds, C.G., Barinaga, C.J. and Udseth, H.R. *Anal. Chem.* **1990**, *62*, 882-899.

**

Note added in proof: Rothchild, K.J. et al., *Proc. Natl. Acad. Sci. U.S.A.* **1989**, 86, 9832-35, used FTIR difference spectroscopy of ^{15}N and deuterated protein to monitor structural changes in Bacteriorhodopoin.

RECEIVED February 6, 1991

Chapter 3

Mechanisms of Inclusion Body Formation

Anna Mitraki, Cameron Haase-Pettingell, and Jonathan King

Department of Biology, Massachusetts Institute of Technology,
Cambridge, MA 02139

The accumulation of newly synthesized polypeptide chains as aggregated inclusion bodies is becoming a serious problem in the recovery of proteins from cloned genes. Studies of both refolding of denatured proteins *in vitro* and of *in vivo* folding and maturation pathways, indicate that aggregates derive from partially folded intermediates in the pathway and not from the native protein. Aggregation is not a function of the solubility and stability properties of the native state, but those of folding intermediates in relation to the environment they are folding in. Ions, cofactors and chaperonins can interact with intermediates and influence the outcome of the folding process. Single amino acid substitutions can suppress aggregation without affecting the activity and stability of the mature protein. Thus, it should be possible to optimize folding pathways with genetic engineering of the intermediates, or alteration of their environment.

The denaturation and aggregation of proteins was, "together with combustion and fermentation, one of the first chemical processes recognized by man" (*1*). Unfortunately this distinctive and characteristic phase change of proteins was very refractory to experiment; the polypeptide chains in the denatured aggregated state were not in equilibrium with soluble species. The particulate nature prevented determination of solution properties and its light scattering properties interfered with spectroscopic techniques. Without the ability to determine the structure in the aggregated state it was natural to consider them structureless. As a result for many decades aggregation was regarded as an annoying complication during *in vitro* denaturation-renaturation studies. As long as refolding was limited to laboratory experiment, the problem could be side-stepped by carrying out refolding experiments at low protein concentrations, where aggregation is minimized (*2*).

The development of methods of cloning foreign genes into heterologous hosts has brought aggregation back as a scientific and practical problem. A common outcome of the expression of cloned genes turned out to be not the native protein, but aggregated states, termed "inclusion bodies". The overexpressed polypeptides

0097–6156/91/0470–0035$06.00/0

fail to reach their native state and instead accumulate in an aggregated, inactive form. Those protein aggregates sediment at low speed, are amorphous and are not surrounded by any kind of membranes, when observed in the electron microscope (3, 4). Initially it was thought that the protein state in those inclusion bodies was a covalently incorrect or damaged state, analogous to the inclusion bodies formed by abnormal proteins in hemoglobin and other diseases (5, 6) or in *E. coli* (7). This model was dismissed with the demonstration, for numerous proteins, following dissociation of the aggregated chains with strong denaturants, the native state could be recovered after return to physiological conditions (reviewed in 8). The requirement for stringent solubilization conditions also indicated that the formation of inclusion bodies was not simply due to overcoming of solubility limits of the protein. With the recognition that inclusion body formation might be a conformational or folding problem, incorrect disulphide bond formation was considered as the likely culprit (9). Though present in inclusion bodies recovered after cell lysis, disulphide bonds are unlikely to be the primary problem. Proteins without disulphide bonds or cysteines still accumulate in inclusion bodies (8, 10).

In 1989 we reviewed the literature on aggregation during refolding *in vitro*, and together with our own studies on aggregation *in vivo*, we concluded that inclusion body formation represented the interaction of folding intermediates, due to partial denaturation or absence of an essential cofactor in the heterologous cytoplasm (11). Since then considerable experimental support has emerged for this view.

Inclusion Bodies and Protein Solubility.

Inclusion body formation and precipitation of native proteins at their isoelectric point or during salting out are two fundamentally different phenomena. In a precipitate, the molecules are held together by interactions between their surfaces. No major conformational changes occur between the protein in the precipitated state and the solution state (12, 13). As a result, the structural and functional properties of the macromolecule can be regained after re-dissolving or readjusting the pH (12, 14). The aggregated state differs sharply from the precipitated state. It is generally not possible to release soluble protein from inclusion bodies by dilution into native buffer conditions (8). Where part of the protein is produced in the active soluble form and the rest as inactive aggregated material, the end state populations are not in equilibrium. Solubilization of inclusion bodies requires disruption of the forces that hold them together by strong denaturants, implying that the interactions between chains in the aggregate share characteristics of the intrachain interactions holding native proteins together. If the denaturant is dialyzed or diluted out without attempting to find refolding conditions, the chains frequently reaggregate. Solubilization of the chains under native conditions requires refolding to the native conformation. However, during *in vitro* refolding, aggregate formation often interferes with the acquisition of the native form (15, 16). Therefore, inclusion body formation *in vivo* and aggregation observed as an off-pathway step during *in vitro* refolding are closely related processes and can be treated in the same conceptual framework.

Solubility and Stability of Folding Intermediates Versus Native Forms

Attempts to empirically correlate inclusion body formation with protein size, sequence, production level and rate have not given any conclusive information (*10, 17*). This presumably reflects the confusion between precipitation processes which depend on properties of the native state, and the aggregation of polypeptide chains newly synthesized with the cell, which is a property of the *in vivo* folding intermediates. During both *in vivo* folding and *in vitro* refolding the unfolded polypeptide chain forms and passes through intermediate conformations (*18, 19, 20*). These intermediate conformations are not just sub-sets of the final native one, but can be different, and thus can have distinct physical properties (*21, 22*). Given this we have to take into account that the properties of folding intermediates cannot be simply deduced from the properties of the native state. At the present time properties of folding intermediates cannot be predicted but must be determined empirically.

In vitro aggregation studies have provided evidence that there is kinetic competition between the folding and aggregation pathways and aggregates form from folding intermediates rather than from native or unfolded proteins (*15, 23*). The role of intermediates during the *in vivo* folding of proteins is also becoming evident from a number of studies (*24, 25*). The ensemble of those studies has been previously reviewed in detail (*11*). Below we will review some evidence suggesting that solubility and stability properties of intermediates, and not those of the native state must be the critical factor in aggregate formation, both *in vitro* and *in vivo*.

A protein for which solubility of different folding conformers has been carefully studied *in vitro* is bovine growth hormone, a monomeric four-helix bundle protein. Using two-step kinetic procedures, David Brems has demonstrated that during unfolding and refolding the native and denatured forms of the protein are soluble or give rise to products that are soluble (*26*). However, a partially folded intermediate can be populated under moderately denaturing conditions which lead to aggregation after transfer to non-denaturing solvents. This study clearly demonstrated that association events between partially folded monomeric intermediates lead to an associated intermediate that is insoluble under native conditions. Aggregation was linearly dependent on the protein concentration present in the initial step of the procedure. Thus, those intermolecular interactions between folding intermediates are a prerequisite for aggregation. Brems et al. (*27*) have proposed that the partially folded monomeric intermediate possess exposed lipophilic faces of the putative amphiphilic third helix. Formation of the associated intermediate subsequently proceeds through intermolecular hydrophobic interactions involving these hydrophobic faces.

Inclusion body formation is often suppressed by lowering the temperature (*28-31*). This has led to the assumption that the aggregation process is due to thermal denaturation of the native form of the protein. Schein (*17*) postulated which proteins that have higher melting points and higher native state stability should not form inclusion bodies. However, from the limited number of *in vivo* aggregation studies, aggregate formation precedes the final maturation steps (in the case that native state and aggregates co-exist) or aggregates form even though the native state has never existed in the cells. This is particularly clear in the case of P22 tailspike protein, which is a highly stable, predominantly beta-sheet homotrimer. This protein is one of the very few procaryotic systems that provides a model for inclusion body formation *in vivo* (*29*). The native form of the protein is highly thermostable (T_m is 88° C) and resistant to SDS, proteases and heat. However, during the *in vivo* folding process, the polypeptide chain passes through intermediate conformations that are sensitive to all those factors (*24*).

The maturation of the wild type polypeptide in the cell proceeds with about 25% efficiency at high temperatures (39° C). The polypeptide chains that do not reach the native form, accumulate in aggregates (*33*). The kinetics of aggregate formation show that aggregates derive from early partially folded intermediates, and not from the native state. In fact, this early intermediate can either enter the productive pathway or form aggregates as seen in Figure 1. This partition is temperature dependent, with the aggregation pathway favored at high temperatures as seen in Figure 2. If chains that have been synthesized at a high temperature are shifted to low temperature early enough, they can reenter the productive pathway (*29, 34*). Furthermore, native tailspikes produced at permissive temperatures *in vivo*, stay native if shifted up to restrictive temperatures, proving that once the native form is attained, its solubility and stability properties are not altered at high temperatures (*35, 36*). Thus, inclusion body formation at high temperatures can be explained by the presence of a thermolabile intermediate in the folding pathway, which can partially denature and give rise to a species that can aggregate. The tailspike example might not be representative, because of the unusual stability of the native protein. We believe this is not an exception to the general rule, but simply offers a technical and conceptual advantage; it is the dramatic difference of physical properties between the native state and intermediates that allows a clear distinction of the phenomena originating from each form.

What is the Conformation of the Chains in the Inclusion Body and the Nature of Forces that Hold Them Together?

Since protein aggregates derive from partially folded intermediates in folding pathways, it is tempting to suggest that they should conserve conformational features of those intermediates. However, due to their physical state, it is very difficult to directly probe the aggregate structure. Rainer Jaenicke and colleagues have shown by circular dichroism measurements, that aggregated chains indeed retain secondary structure (*15*) as seen in Figure 3. The same conclusion was reached by George Thomas and co-workers, from Raman spectroscopy data of *in vitro* produced tailspike aggregates (*37*). In another study, Michel Goldberg has used monoclonal antibodies to characterize the state of renatured forms of β2 tryptophan synthetase subunits, after acid denaturation (*38*). This protein is one of the systems for which the relationship between intermediates and the aggregation process has been explicitly addressed (*39*). Upon renaturation, three forms of the protein are obtained: one soluble, identical to the native β2, insoluble high molecular weight aggregates and soluble aggregates of a low polymerization degree. The latter form of aggregates have the same immunoreactivity as native β2 with two monoclonal antibodies. However this species is not recognized by other three antibodies, indicating that although non-native, aggregates bear "native-like" features.

From the overall number of *in vivo* and *in vitro* studies, we know that the intermolecular forces involved in aggregate formation are non-covalent and are

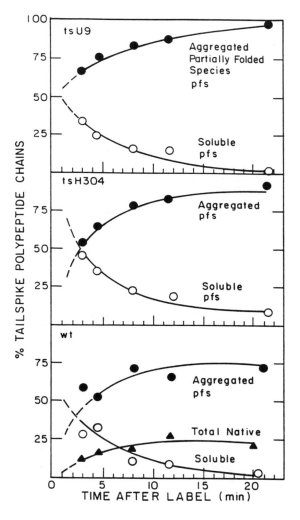

Figure 1: Time course of intracellular aggregation of wild type and *tsf* mutant forms of P22 tailspike polypeptide chains. Infected cells were incubated with ^{14}C amino acids at 39° C. Samples were chilled and harvested at different times after label. After lysis by freezing and thawing the lysate was fractionated into pellet and supernatant. They were mixed with sample buffer and electrophoresed through an SDS polyacrylamide gel (*68*). Each value represents the percent of total labelled tailspike polypeptide chains present in the sample. (▲) native tailspikes, (○) folding intermediates, (●), aggregated species (derived from intermediates). (Reproduced with permission from reference 29. Copyright 1988 The American Society for Biochemistry & Molecular Biology.)

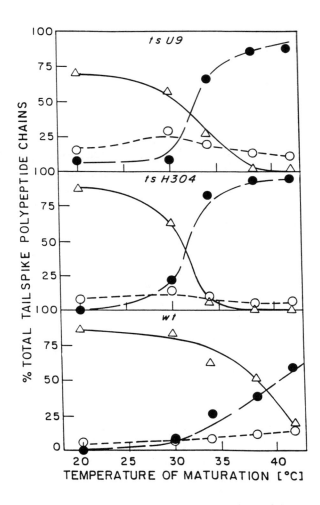

Figure 2: Effect of temperature on the aggregation of intermediates for wild type and *tsf* mutants of the P22 tailspike. The experimental protocol is the same as in figure 1 except that after labelling, the polypeptide chains were allowed to mature for one hour at different temperatures. The percentage of each species at the end of maturation is plotted as a function of maturation temperature. (△) native tailspikes, (○) folding intermediates, (●) aggregated species. (Reproduced with permission from reference 29. Copyright 1988 The American Society for Biochemistry & Molecular Biology.)

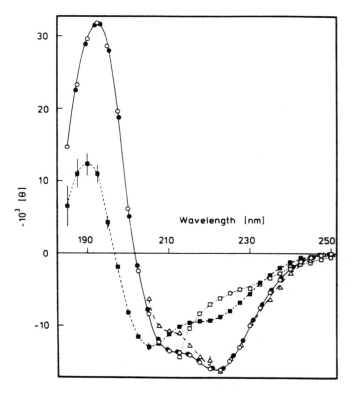

Figure 3: Far-UV circular dichroism spectra of native, denatured, and aggregated forms of LDH-M_4 (tetrameric porcine lactic dehydrogenase from skeletal muscle). (●) native and (O) renatured LDH-M_4 at a concentration of 1.5 mg/ml. (■) LDH-M_4 denatured at pH 2 or at (□) 6 M guanidine hydrochloride , same concentration; (△) aggregated LDH-M_4 dispersed by low-energy sonication at a concentration of 0.05 mg/ml (determined gravimetrically from the dry weight at 105° C). From Zettlemeissl et al, (*15*) reproduced with permission.

predominantly hydrophobic interactions (11). In every case where partially folded intermediates that carry exposed hydrophobic structural elements can be populated, there is a high probability of incorrect intermolecular association. For example this can happen when β strands get exposed in an α/β protein (phosphoglycerate kinase, rhodanese), and in predominantly β protein, (tailspike, γ crystallin) or in the case of exposure of the lipophilic face of an amphiphilic helix (bovine growth hormone) (11, 40, 41). The weakening of hydrophobic forces at low temperature can then account for aggregate suppression at low temperatures, both *in vitro* and *in vivo*.

Is Aggregate Formation Specific Or Not?

In order to answer this question, one needs to explicitly investigate the structure of the partially folded intermediates, as well as their subsequent polymerization mechanism. Information on the nature of the steps that precede the formation of large aggregates is limited for the moment. In a recent study, Cleland and Wang have directly demonstrated, using quasi-elastic light scattering, that dimeric and trimeric forms of an early monomeric partially folded intermediate are the precursors to large aggregate formation in the case of carbonic anhydrase (42 and this volume). In addition the bovine growth hormone study has showed that the critical events determining the aggregation lie at the step that populates the associated folding intermediate (26). When introduced into this initial step, a number of peptide fragments that correspond to certain parts of the critical amphiphilic helix can inhibit aggregation (26) as observed in Figure 4. Inhibition of aggregation can be explained if the helical peptide can stereospecifically interact with the exposed hydrophobic part of the corresponding helix carried by the monomeric intermediate, thus preventing formation of the associated intermediate. Only peptides carrying the C-terminal part of the third helix can play this inhibitory role, indicating that association can involve distinct structural elements. Thus, one can hypothesize that if aggregation proceeds according to a mechanism which involves specific interaction of structural elements, then certain positions in the polypeptide chain will carry information relevant to the process. If this holds true, then single amino acid substitutions could either raise or lower the aggregation yield. Moreover, since the stability of the native form is not the factor determining aggregate formation, one could predict that this kind of substitution will not affect the properties of the native protein.

Mutational Studies of the Aggregation Phenomena.

There are a limited number of examples of mutations that can influence aggregate formation. Krueger et al. (43) have found that single amino acid substitutions in overexpressed signal transduction proteins that control motility in *E. coli*, cause inclusion body formation. For bovine growth hormone, substitution of a lysine for a leucine extends the hydrophobic face of the amphiphilic helix and therefore should enhance the forces that stabilize the associated intermediate (44). The conformation of the native form of this mutant was indistinguishable from the wild-type protein, as judged by different conformational criteria. However, the refolding of the mutant protein was slower and was accompanied by enhanced aggregation, due to the stabilization of the associated intermediate.

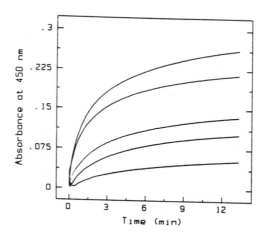

Figure 4: Inhibition of aggregation of bovine growth hormone (BGH) by a peptide fragment comprising residues 96-133 of the hormone. A 1.75 mg/ml solution of the protein was initially incubated at 3.5 M guanidine hydrochloride (conditions that populate the associated intermediate) and subsequently was diluted at 0.18 mg/ml and 0.8 M guanidine hydrochloride (conditions that induce aggregation). Turbidity was monitored by the absorbance at 450 nm. The top curve represents the kinetics of turbidity formation in the absence of the peptide, and in descending order from the top, formation of turbidity in the presence of 1, 3, 5, 7-fold molar excess of the peptide. The peptide was present from the initial conditions that populate the intermediate. (3.5 M guanidine hydrochloride). From Brems, (26) reproduced with permission.

For the tailspike system, a class of mutants called temperature sensitive for folding, (tsf) have been isolated which alter the folding pathway without influencing the properties of the native form (45). The tsf mutant polypeptide chains fail to reach the native form at high temperatures in vivo, but at permissive temperatures they form the native protein. This native form of the mutant protein is as thermostable as the wild-type protein (35, 36). The tsf polypeptide chains synthesized at high temperature form an early folding intermediate which aggregates. In this case, the native form has never existed in the cells, confirming that the aggregation pathway does not pass through the mature form (29). The biochemical analysis of the purified native forms of the mutant proteins, formed at permissive temperatures, has shown that their structural and functional properties are not significantly altered with respect to the wild-type form (36, 46-48). Thus, these single amino acid changes do not act at the level of the mature protein, but can further destabilize an already thermolabile intermediate in the folding pathway, or alternatively, speed up the off-pathway step.

The defects of tsf mutants can be alleviated by another class of mutations, second-site suppressors (49). Two such mutations were able to suppress defects associated with a number of tsf mutations spanning a 200-residue distance in the linear sequence. This dispersion in combination with the fact that there was no obvious pattern relating the side chains at the suppressor sites to those at the starting mutant sites, suggested that the suppression must follow a global or general mechanism (50). Analysis of the in vivo maturation of the chains carrying both the tsf and the suppressor mutations indicates that the fraction of the aggregated chains is reduced at high temperatures. Furthermore, the chains carrying only the suppressor mutation mature more efficiently than wild-type chains at restrictive temperature, again with a concomitant decrease in the aggregated chains. The activity and stability of the mature suppressor proteins are not altered (unpublished results). Thus, it seems that those positions indeed carry information relevant to the folding pathway that can be important for the aggregation phenomena. The investigation of the effect of both tsf and suppressor mutations on the in vitro refolding of the purified proteins is currently under way. We believe that their further study will provide a better understanding of the aggregation mechanisms.

The Role of Environmental Conditions in Aggregate Formation.

The necessity of a proper environment for attainment of the functional protein state has long been recognized (51). The most fundamental environmental factors can be designated as temperature, pH, ionic strength and redox potential. Moreover, ligands and cofactors have been proven to interact with intermediates, and influence the outcome of the folding process (reviewed in 11, 40). It is evident that by overexpressing a protein in a heterologous cytoplasm, the difference in environment as well as in ionic and cofactor composition can lead to failure of the folding intermediates to proceed through the folding pathway.

The fact that inclusion bodies are associated with the overexpression of cytoplasmic proteins, has led to the assumption that aggregation can be circumvented by secretion. However, Georgiou et al. have shown that overexpression of a periplasmic protein native to E. coli such as β-lactamase can also result in inclusion bodies (52). The formation of inclusion bodies is inhibited by the presence of non-metabolizable sugars such as sucrose and raffinose (53).

Chaperonins

An exponentially growing number of studies has revealed that "helper proteins" may be required for the attainment of the final mature conformation. Such proteins include protein disulphide isomerase, proline cis-trans isomerase and molecular chaperonins (54-56). The latter class was described as a class of homologous proteins in bacteria, plants and animals required for the proper assembly of oligomeric proteins (57). Bochkareva et al. (58) have subsequently demonstrated transient association of newly synthesized pre-β lactamase and chloramphenicol transacetylase chains with the GroEL chaperone, suggesting that the chaperonin role starts at earlier steps of folding than final assembly steps. GroEL/ES chaperonin proteins are required for the assembly of procaryotic ribulose bisphosphate carboxylase (Rubisco) oligomers in *E. coli* (59). Using purified proteins, George Lorimer and coworkers have demonstrated that chaperonin proteins can bind to partially folded intermediates of Rubisco that are prone to aggregation and assist them through the productive pathway *in vitro*. Under those experimental conditions, absence of chaperonin results in total failure of renaturation (60). They have also convincingly demonstrated that the chaperonins do not interact with the native form of the protein, and they are unable to rescue Rubisco chains that have already aggregated. It is interesting to note that although starting from a different point, those studies provide evidence that the critical steps in aggregate formation lie at the level of folding intermediates.

For pre-secretory proteins, helper proteins function in maintaining an "unfolded state" necessary for translocation (61, 62). In a recent study, Bill Wickner and co-workers, using purified proteins, reported that an *E. coli* outer membrane precursor protein, proOmpA, is competent for *in vitro* translocation immediately after dilution from denaturing urea concentrations, but loses translocation competence as a function of time in the renaturing mixture (63). The authors determined that this loss of translocation competence is due to aggregation, and SecB binding inhibits aggregation (64). The translocation competent form that interacts with SecB already possesses secondary and tertiary structure, so the SecB role might lie in preventing aggregation of proOmpA folding intermediates, rather than simply maintaining an "unfolded state".

Thus, the common factor underlying aggregate formation is failure of partially folded intermediates to correctly proceed through the productive pathway. This may result from their partial intracellular denaturation, due for example to temperature, or lack of proper environmental conditions. Absence of necessary cofactors and helper proteins, or failure to interact with them at a critical stage may also result in aggregation of folding intermediates as depicted in Figure 5.

From Theory to Praxis: Strategies for Native Protein Recovery.

The understanding of protein folding pathways and stabilities of folding intermediates can lead to rational design of folding protocols. Although the choice of the proper conditions requires information on the folding pathway for each particular system, some general rules can apply. Temperature is a factor of quite

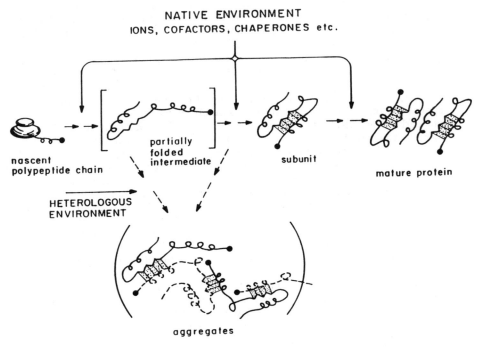

NATIVE ENVIRONMENT
IONS, COFACTORS, CHAPERONES etc.

nascent
polypeptide chain

partially
folded
intermediate

subunit

mature protein

HETEROLOGOUS
ENVIRONMENT

aggregates

Figure 5: *In vivo* hypothetical folding and maturation pathway for a dimeric protein. The outcome of the pathway depends on the existence of the proper environmental conditions. (temperature, pH, ionic strength, cofactors, chaperonins etc.). A heterologous environment can influence the fate of the partially folded intermediates towards the aggregation pathway. The dotted areas denote intrachain interactions in the native form of the protein. Interactions between chains in the aggregate share characteristics of those intrachain interactions. This schematic representation tries to emphasize the aggregation steps, and therefore does not represent conformational refinements and isomerization steps that may occur as late events in the folding pathway. (Reproduced with permission from reference 11. Copyright 1988 Bio/Technology.)

general importance for the intracellular stability of folding intermediates. Suppression of inclusion body formation at lower culture temperatures has been described for a number of systems (*28-31*). When production of protein in soluble form *in vivo* is desired, lower temperature culture conditions (around 30° C) may be advantageous. The use of fusion polypeptides might be a source of concern since hydrophobic fusion sequences can contribute to inclusion body formation (*32*). When a ligand, cofactor, or ion is necessary for correct maturation, it must be provided if is not present in sufficient amounts in the host environment. Coexpressing a molecular chaperone may become a common practice (*59*).

In the case where inclusion bodies still form *in vivo* process, different techniques can help to achieve the active protein form after *in vitro* denaturation-renaturation. For a review of the classical choice of unfolding and refolding conditions, see Jaenicke and Rudolph (*65*). If the classical refolding treatments still fail to yield the native form, there is the possibility of detergent-assisted or cosolvent-assisted refolding. Using a non-denaturing detergent (lauryl maltoside), Tandon and Horowitz (*66*) have successfully renatured rhodanese after guanidine hydrochloride denaturation. Lauryl maltoside binds to exposed hydrophobic surfaces of folding intermediates, preventing aggregation. Using polyethylene glycol, which is a cosolvent that binds to nonpolar regions of an early intermediate and is subsequently excluded from more compact intermediate forms and from the native protein, Cleland and Wang recently prevented aggregation of carbonic anhydrase (*67* and this volume). Finally, our work has demonstrated that single amino acid substitutions can specifically carry information that suppresses the aggregation pathway, without affecting the activity and stability of the mature protein. This suggests that it is possible to optimize folding pathways without altering the desired properties of the final native form. Thus, the possibility of suppressing aggregation using genetic engineering can be envisaged.

Literature Cited

1.   Anson, M. L.; Mirsky, A.E. *J. of Phys. Chem.* **1931** *35*, 185-193.
2.   Anfinsen, C.B. and Haber, E. *J. Biol. Chem.* **1961** *236*, 1361-1363.
3.   Schoner, E. G.; Ellis, L. F.; Schoner, B. E. *Bio/Technology* **1985** *3*, 151-154.
4.   Taylor, G.; Hoare, M.; Gray, D. R.; Marston, F.A.O. *Bio/Technology* **1986** *4*, 553-557.
5.   Carrell, R.W.; Lehmann, H.; Hutchison, H.E. *Nature* **1966** *210*, 915-916.
6.   Schneider, R.G.; Ueda, S.; Alperin, J.B.; Brimhall, B.; Jones, R.T. *The New Eng. J. of Med.* **1969** *280*, 739-745.
7.   Prouty, W.F.; Goldberg, A.L. *Nature New. Biol. (London)* **1972** *240*, 147-150.
8.   Marston, F. A. O. *Biochem. J.* **1986** *240* 1-12.
9.   Schoemaker, J.M.; Brasnett, A.H.; Marston, F.A.O. *EMBO J.* **1985** *4*, 775-780.
10. Kane J.F.; Hartley D.L. *Trends in Biotech.* **1988** *6*, 95-101.
11. Mitraki A.; King J. *Bio/Technology* **1989** *7*, 690-697.
12. Scopes, R. *Protein purification.* Springer-Verlag, New York, N. Y., 1982 pp 40-52.
13. Volkin, D. B.; Klibanov, A. M. In *Protein function. A practical approach.* Creighton, T., Ed.; IRL press, Oxford, 1988.

14. Schein, C. H. *Bio/Technology* **1990** *8*, 308-317.
15. Zettlmeissl, G.; Rudolph, R.; Jaenicke, R. *Biochemistry* **1979** *18*, 5567-5571.
16. Rudolph, R.; Zettlmeissl, G.; Jaenicke, R. *Biochemistry* **1979** *18*, 5572-5575.
17. Schein, C. H. *Bio/Technology* **1989** *7*, 1141-1149.
18. Kim, P.S.; Baldwin, R.L. *Ann. Rev. Biochem.* **1982** *51*, 459-489.
19. Kim, P.S.; Baldwin, R. L. *Ann. Rev. Biochem.* **1990** *59*, 631-660.
20. Goldenberg, D.P.; Smith, D.H.; King, J. *Nat. Acad. Sci. USA* **1983** *80*, 7060-7064.
21. Creighton, T.E. *Prog. Biophys. Mol. Biol.* **1978** *33*, 231-298.
22. Ptitsyn O.B. *J. of Prot. Chem.* **1987** *6*, 273-293.
23. London J.; Skrzynia C.; Goldberg M. 1974. *Eur. J. Biochem.* **1974** *47*, 409-415.
24. Goldenberg, D.; King, J. *Proc. Natl. Acad. Sci. USA* **1982** *79*, 3403-3407.
25. Hurtley, S.; Helenius A. *Annu. Rev. Cell Biol.* **1989** *5*, 277-307.
26. Brems D. N. *Biochemistry* **1988** *27*, 4541-4546.
27. Brems, D. N.; Plaisted, S. M.; Kauffman, E. W.; Havel, H. A. *Biochemistry* **1986** *25*, 6539-6543.
28. Mizukami, T.; Komatsu, Y.; Hosoi, N.; Hoh, S.; Oka, T. *Biotech. Letters* **1986** *8*, 605-610.
29. Haase-Pettingell, C.; King J. *J. Biol. Chem.* **1988** *263*, 4977-4983.
30. Schein, C.H.; Noteborn, M.H.M. *Bio/Technology* **1988** *6*, 291-294.
31. Takagi, H.; Morinaga, Y.; Tsuchiya, M.; Ikemura, H.; Inouye, M. *Bio/Technology* **1988** *6*, 948-950.
32. Lee, S.C.; Choi, Y.C.; Yu, M-H. *Eur. J. Biochem.* **1990** *187*, 417-424.
33. Goldenberg, D. P.; Berget, P. B.; King, J. *J. Biol. Chem.* **1982** *257*, 7864-7871.
34. Smith, D.H.; King, J. *J. Mol. Biol.* **1981** *145*, 653-676.
35. Goldenberg, D.; King, J. *J. Mol. Biol.* **1981** *145*, 633-651.
36. Sturtevant, J.; Yu, M-h; Haase-Pettingell, C.; King, J. *J. Biol. Chem.* **1989** *264*, 10693-10698.
37. Sargent, D.; Benevides, J.M.; Yu, M-h; King, J.; Thomas, Jr., G.J. *J. Mol. Biol.* **1988** *199*, 491-502.
38. Murry-Brelier, A.; Goldberg, M.E. *Biochimie* **1989** *71*, 533-543.
39. Goldberg M.; Zetina C. In *Protein Folding*; R. Jaenicke Ed. Elsiever, North Holland Biomedical Press. 1980 pp. 469-484.
40. Jaenicke, R. *Prog. in Biophys. and Mol. Biol.* **1987** *49*, 117-237.
41. Mitraki A.; Betton J.-M.; Desmadril M.; Yon J. *Eur. J. Biochem.* **1987** *163*, 29-34.
42. Cleland, J.L.; Wang, D. *Biochemistry* **1990a** *29*, 11072-11078.
43. Krueger, J.K.; Stock, A.M.; Schutt, C.; Stock, J.B. In *Protein folding. Dechiphering the second half of the genetic code;* Gierasch, L. and King, J. Eds. A. A. A. S. Washington, D.C. 1990 pp.137-142.
44. Brems, D.N.; Plaisted, S.M.; Havel, H.A.; Tomich, C-S.C. *Proc. Natl. Acad. USA* **1988** *85*, 3367-3371.
45. King, J.; Fane, B.; Haase-Pettingell, C.; Mitraki, A.; Villafane, R.; Yu, M-h. In *Protein Folding, Deciphering the second half of the genetic code*; Gierasch, L.M. and King, J., Eds. A.A.A.S., Washington D. C. 1990 pp. 225-239.
46. Yu, M.-H.; King, J. *Proc. Natl. Acad. Sci. USA.* **1984** *81*, 6584-6588.

47. Yu, M.-H.; King, J. *J. Biol. Chem.* **1988** *263*, 1424-1431.
48. Thomas, G. J. Jr.; Becka, R.; Sargent, D.; Y, M-H.; King J. *Biochemistry* **1990** *29*, 4181-4187.
49. Fane, B. and King, J. *Genetics* **1991** *127*, 263-277.
50. Fane, B.; Villafane, R.; Mitraki, A.; King, J. *J. Biol. Chem.* subbmitted
51. Anfinsen, C.B. *Science* **1973** *181*, 223-230.
52. Georgiou, G.; Telford, J.N.; Shuler, M.; Wilson, D.B. *Applied and Environmental Microbiology* **1986** *52*, 1157-1161.
53. Bowden, G. A.; Georgiou, G. *Biotechnology progress* **1988** *4*, 97-101.
54. Pelham, H. R. B. *Cell* **1986** *46*, 959-961.
55. Freedman, R. *Cell* **1989** *57*, 1069-1072.
56. Fischer, G.; Schmidt, F.X. *Biochemistry* **1990** *29*, 2205-2212.
57. Hemmingsen, S. M.; Woolford, C.; van der Vries, S. M.; Tilly, K.; Dennis, D. T.; Georgopoulos, C. P.; Hendrix, R. W.; Ellis, R. J. *Nature* **1988** *333*, 330-334.
58. Bochkareva, E. S.; Lissin, A. S.; Girshovich, A. S. *Nature* **1988** *336*, 254-257.
59. Goloubinoff, B.; Gatenby, A.A; Lorimer, G. *Nature* **1989a** *337*, 44-47.
60. Goloubinoff, P.; Christeller, J.T.; Gatenby, A. A.; Lorimer, G. H. *Nature* **1989b** *342*, 884-889.
61. Randall, L. L.; Hardy, S. J. S.; Thom, J. A. *Ann. Rev. Microbiol.* **1987** *41*, 507-541.
62. Meyer, D. I. *Trends Biochem. Sci.* **1988** *13*, 471-474.
63. Lecker, S.; Lill, R.; Ziegelhoffer, T.; Georgopoulos, C.; Bassford, P.J.; Cumamoto, C.A.; Wickner, W. *EMBO J.* **1989** *8*, 2703-2709.
64. Lecker, S.H.; Driessen, A.J.M.; Wickner, W. *EMBO J.* **1990** *9*, 2309-2314.
65. Jaenicke, R.; Rudolph, R. In *Protein structure. A practical approach;* Creighton, T. Ed. IRL press, Oxford. 1990 pp 191-223.
66. Tandon S.; Horowitz P. *J. Biol. Chem.* **1986** *261*, 15615-15681.
67. Cleland, J.L.; Wang, D.I.C. *Bio/Technology* **1990b** *8*, 1274-1278.
68. King, J.; Laemmli, U. K. *J. Mol. Biol.* **1971** *62*, 465-477.

RECEIVED March 8, 1991

Chapter 4

Structure and Stability of Cytochrome c Folding Intermediates

Gülnur A. Elöve and Heinrich Roder

Department of Biochemistry and Biophysics, University of Pennsylvania,
Philadelphia, PA 19104–6059

Hydrogen exchange labeling and nuclear magnetic resonance
(NMR) approaches were used to elucidate the folding mechanism
of horse cytochrome c (cyt c). The development of hydrogen
bonded structure during refolding was observed by pulse labeling at
variable refolding times, and the degree of protection from NH
exchange was probed by systematic variation of the labeling pH.
The results show that the folding reaction involves both sequential
and parallel pathways. About 50% of the molecules follow a
sequential pathway where a partially folded intermediate is formed
in a 20 ms folding phase. This intermediate has two native-like
helices near the chain termini, but lacks stable H-bonded structure
in other parts of the molecule. Amide sites on either helix are not
only protected at the same rate, but they also exhibit the same
degree of protection, confirming previous evidence that association
of the N- and C-terminal helices is an important early event in cyt c
folding. Subsequent folding events on the 100 ms time scale involve
replacement of a non-native histidine heme ligand by the native
methionine ligand. In addition, there is evidence for a minor
species of very rapidly folding molecules that form native-like
structure within the mixing dead-time, as well as slow-folding forms
that take several seconds to fold.

Much has been learned in recent years about the folding process of small globular
proteins, especially through application of new experimental approaches such as
two dimensional nuclear magnetic resonance (2D NMR) and site-directed
mutagenesis (1-4). However, we are still far from a detailed understanding of how
the three-dimensional structure of a protein is encoded in its amino acid
sequence. Once mainly of academic interest, the protein folding problem has
found new motivation from recent developments in protein engineering and
biotechnology.
   For a mechanistic understanding of protein folding, it is necessary to
determine the sequence of intermediate states between the fully unfolded and
fully folded forms of a polypeptide chain (i.e. the folding pathway) and to
describe the major intermediates in structural and kinetic terms. One of the

0097–6156/91/0470–0050$06.00/0

challenges encountered in the effort to characterize folding intermediates is the cooperativity of the folding transition, which makes it difficult to study partially folded intermediates at equilibrium. Folding intermediates are more likely to become populated under non-equilibrium conditions (*1*), but kinetic intermediates are usually too short lived to be studied directly by structurally powerful, but intrinsically slow techniques such as NMR. The limited time resolution of NMR can be overcome by combining it with hydrogen exchange labeling and rapid mixing methods. The feasibility of this approach was first demonstrated in a study of early folding events in bovine pancreatic trypsin inhibitor (BPTI) (*5*), using one-dimensional NMR in conjunction with a competition method (*6*). Later, a more powerful pulse labeling method developed independently in this laboratory and that of Robert Baldwin, has been applied to study the folding kinetics of cytochrome c (*7*), ribonuclease A (*8,9*), and more recently to barnase (*10*) and ubiquitin (*Briggs and Roder, in preparation*).

The hydrogen exchange labeling methods rely on backbone amide protons as probes for the formation of hydrogen bonded structure during refolding and uses ^{1}H NMR to observe protons that have been protected from exchange in the refolded protein. The intrinsic exchange behavior for the amide protons of an unstructured polypeptide chain is well known from studies of model peptides and unfolded proteins (*11,12*). Since the exchange reaction is dominated by base catalysis above pH 3, pH can be used to vary the rate over a time scale ranging from tens of minutes (at pH 3) to milliseconds (at pH>9). On the other hand, the exchange rates for hydrogen bonded amide protons in folded proteins can be retarded by many orders of magnitude (*13-17*). The large majority of these slowly exchanging amide protons in several proteins were found to be involved in internal hydrogen bonds (*15,16*), suggesting that H-bond formation is the dominant factor in retarding the rate of NH exchange in native proteins. For example, the 42 most slowly exchanging amide protons in oxidized cyt c (*17*) are all involved in crystallographically defined hydrogen bonds. The exchange rates of these protons are retarded by factors of 10^4 to $>10^8$ relative to model peptides, which makes them sufficiently long-lived for 2D NMR analysis.

In a typical pulse labeling experiment, the protein is unfolded in D_2O and refolding is initiated by rapid dilution of a denaturing agent. The time course of structure formation is probed by briefly exposing the partially refolded protein to H_2O under rapid exchange conditions (basic pH) so that amide sites in unstructured parts of the polypeptide chain become completely protonated while sites involved in H-bonded structure remain deuterated. After the labeling pulse, the protein is allowed to refold at mildly acidic pH where exchange is slow and the labels become trapped in the refolded protein. The proton labeling pattern imprinted by the pulse is then analyzed by using 2D NMR to measure the amount of proton label trapped at protected amide sites (e.g. at the 42 slowly exchanging NH sites of cyt c). In a series of such experiments at different refolding times, one can thus measure the time course of structure formation in terms of individual hydrogen bonds for those regions of the protein where NH sites become protected from exchange upon folding.

This chapter describes the application of hydrogen exchange labeling and 2D NMR methods to study the folding pathway of oxidized horse heart cyt c. The results illustrate the capability of this approach to provide a detailed picture of the structural events that take place during folding. We also show that additional information on the mechanism of folding and the stability of folding intermediates can be obtained by systematic variation of the exchange conditions at certain points in folding. Cytochrome c is among the best characterized proteins in terms of structure, folding and NMR, and is thus an ideal model protein for such studies. In particular, high-resolution x-ray structures have been reported for several bacterial and mitochondrial c-type cytochromes (*18-20*), including horse heart cyt c (*21*). Cyt c has also been the subject of numerous studies on thermodynamic and kinetic aspects of folding (*22-31*). Moreover, essentially complete ^{1}H NMR assignments have recently been reported for both oxidation states of horse cyt c (*32-33*).

Pulse Labeling Strategy

The rapid mixing procedures of a typical pulse labeling experiment are illustrated in Figure 1, using our previous studies on oxidized cyt c (*7*) as an example. The protein is initially unfolded in a D_2O solution containing 4.5 M GuHCl and is allowed to reach equilibrium so that all the exchangeable amide protons are replaced by deuterons. In the first mixing chamber, refolding is initiated by dilution of the denaturant with excess D_2O buffer or H_2O under slow-exchange conditions (e.g. pH 6 where NH exchange is slow compared to the rate of folding). After variable refolding periods (t_f), deuterium-hydrogen exchange is started by mixing the partially folded protein with excess H_2O buffer at pH 9.5 (second mixer). The labeling conditions are chosen such that exposed amide sites become protonated within a few milliseconds while sites protected in hydrogen bonded structure remain deuterated. After about 50 ms, the exchange reaction is quenched by injecting the mixture into a cold buffer solution at pH 5. Under these conditions, cyt c refolds rapidly compared to further NH exchange so that many of the proton labels introduced by the pulse become trapped in the refolded protein. The protein is then concentrated and 2D J-correlated NMR spectra (COSY) are recorded as previously described (*7*).

Proton occupancies at individual NH sites are determined by normalizing the intensities of resolved NH-CαH COSY cross peaks (or, in some cases, resolved NH resonances in the 1D NMR spectrum), using the spectrum of a fully labeled control sample as a reference. A series of pulse labeling experiments at different refolding times provides kinetic curves that reflect the degree of protection from exchange at many NH sites throughout the protein. In the interpretation of pulse labeling data, it is important to keep the following points in mind. (*i*) For an amide group to remain deuterated during the 50 ms labeling pulse, its exchange reaction must be retarded by at least 50-fold compared to its intrinsic exchange rate, which is about 10^3 s^{-1} at pH 9.5. Therefore, only relatively stable structure gives rise to measurable protection. (*ii*) Amide sites protected to this extent are most likely involved in H-bonded interactions. However, their H-bond acceptors cannot be identified directly. Nevertheless, examination of the patterns of protection for different amide sites in comparison with the native structure can lead to a qualitative picture of the secondary structure present in a folding intermediate. (*iii*) The method can only monitor amide sites that are protected in the native state. The transient formation of a non-native H-bond during folding could not be detected if this site becomes exposed in the folded state. However, the presence of non-native structural elements in a folding intermediate could still be deduced from the patterns of protection. For example, if a segment of the chain is part of a β-sheet early in folding, but forms an α-helix in the native structure, one would expect protection of every other amide site at an early phase and complete protection of all helical NH sites after refolding. On the other hand, if the structure formed early in folding is a native-like α-helix, all the helical amide sites are protected rapidly.

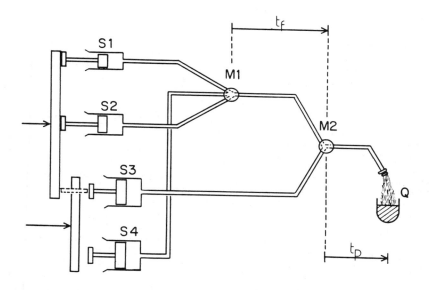

Figure 1.   Schematic diagram of the rapid mixing apparatus used for pulse labeling experiments at variable refolding times (t_f = 3 ms - 10 s) and labeling pulse (t_p = 50 ms):
S1: Horse heart cytochrome c dissolved in 4.5 M GuHCl in D_2O, pH 6.0.
S2: Refolding buffer (100 mM Acetate, pH 6.0).
S3: Pulse buffer (100 mM Glycine, pH 9.5).
Q: Quench buffer (300 mM Citrate, 50 mM Ascorbate, pH 5.3).

Structural and Kinetic Patterns

The time course of protection for a selection of amide protons is shown in Figure 2, using a logarithmic scale for the refolding time. Although different relative amplitudes are observed for different amide protons, one can recognize three common kinetic phases: a fast phase with a time constant of about 20 ms, an intermediate phase of a few hundred milliseconds and a slow phase with a time constant of 5 to 10 sec. Figure 3 shows that the folding kinetics observed by tryptophan fluorescence is also characterized by three kinetic phases with similar time constants, but different relative amplitudes compared to the pulse labeling results. The fluorescence signal is due to a single tryptophan residue, Trp 59, which becomes quenched upon refolding via energy transfer with the heme group. The fact that different experimental probes (42 NH sites and one fluorescence probe) exhibit distinct kinetic behavior demonstrates that partially folded intermediate states are populated on the folding pathway of cyt c.

The majority of amide probes for which the kinetics of protection was measured fall into one of two distinct classes. (i) For a group of protons located near the chain termini (e.g. Val 11, Cys 14, Leu 98, Lys 100), the occupancy drops from 1 to about 0.4 in the first 30 ms of refolding (Figure 2), indicating that they become 60% protected in the fast folding phase. In the intermediate phase, they acquire an additional 20% of protection and become fully protected from exchange in the slow phase. This type of behavior is observed for most of the H-bonded amide protons near the N-terminus up to residue 18 and the C-terminal segment between residues 91 and 100. (ii) For most of the remaining slowly exchanging amide protons, the occupancy remains nearly constant between 0.75 and 0.85 out to about 100 ms, indicating that these sites remain largely accessible for exchange in the early folding phase. The occupancy decreases to about 0.5 in the intermediate phase and drops to zero in the slow folding phase.

In Figure 4, the proton occupancies measured at the end of the fast phase ($t_f = 30$ ms) are mapped onto the native cyt c structure. The rapidly protected protons (class 1) with occupancies below 0.6 are all localized in the contact region between the N- and C-terminal helices, while the more accessible class 2 protons are distributed throughout the rest of the structure. Class 2 protons with occupancies over 0.8 include residues in other helical segments of the native structure (the 60's helix and a short 70's helix) as well as amide sites involved in non-helical H-bonds (reverse turns and irregular tertiary H-bonds).

The labeling pattern seen in Figure 4 indicates that the formation of stable H-bonded structure in the fast folding phase is limited to the N-terminal α-helix (N-helix), the NH of His 18 and the C-terminal α-helix (C-helix). That the structure in these regions corresponds to native-like helices is indicated by the observation of continuous segments of protected protons which is inconsistent with any other type of structure. Furthermore, the following evidence strongly suggests that in the early folding intermediate the two helices already form a hydrophobic helix-helix contact similar to the helix pairing site of the native structure: (i) The time course of protection is virtually identical for amide protons on either helix (Figure 2), indicating simultaneous formation of both helices. (ii) Isolated helices are known to be only marginally stable (*e.g. 35, 36*), and one would not expect strongly protected amide protons without stabilizing tertiary contacts. (iii) The pulse variation results presented below show that protons on either side of the helix-helix interface are protected to the same degree, suggesting again that the stability of the helices depends on their mutual stabilization.

The high labeling levels observed outside the N- and C-terminal regions indicate that the intervening segment of the polypeptide chain (residues 19

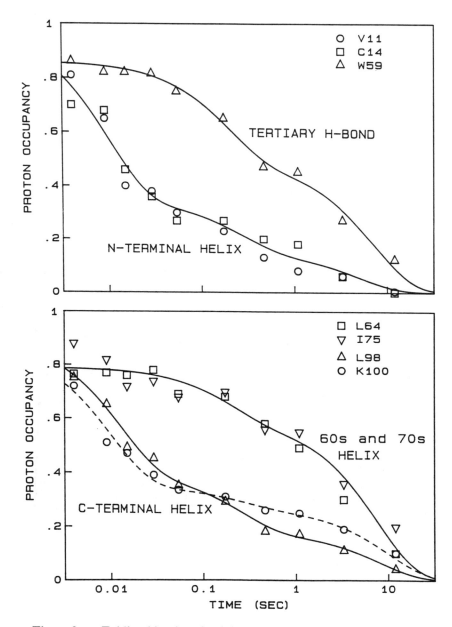

Figure 2.   Folding kinetics of oxidized cyt c at pH 6.0, 10°C, observed by pulse labeling and 2D NMR, using the protocol outlined in Fig. 1. Results are shown for representative amide protons, including examples from the three major helices of cyt c and the indole NH of Trp 59 which forms a tertiary H-bond with a heme propionate side chain. (Reproduced with permission from ref. 7. Copyright 1988 Macmillan Magazines Ltd.)

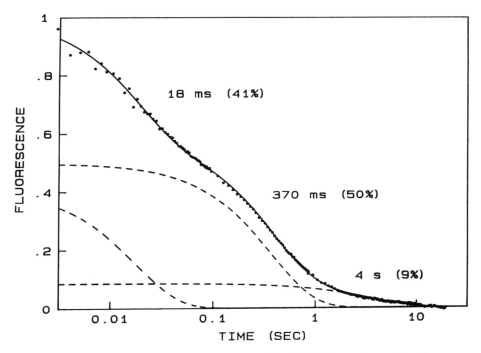

Figure 3.   Folding kinetics of cyt c at pH 6.2, 10°C, monitored by the tryptophan fluorescence (350 nm) on a Hi-Tech stopped-flow apparatus (excitation at 280 nm). Oxidized cyt c was unfolded in 4.2 M GuHCl, pH 6.0 and refolded at pH 6.2 in the presence of 0.7 M GuHCl. (Reproduced with permission from ref. 7. Copyright 1988 Macmillan Magazines Ltd.)

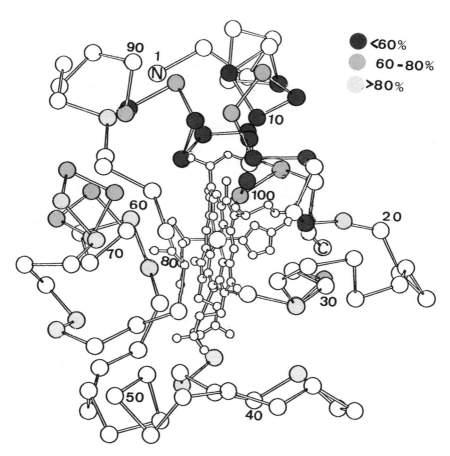

Figure 4.   NH exchange labeling pattern observed in pulse labeling
experiments at t_f = 30 ms (c.f. Fig. 2) coded into the native cyt c
structure. The shading reflects the proton occupancies for 35 NH sites
determined by 2D NMR as described in the text.

through 90) does not contain stable H-bonded structure at this early stage in folding. The lack of strongly protected amide sites is particularly striking for the 60's and 70's region which contains nine slowly exchanging amide protons forming helical H-bonds in the native structure. In the early intermediate, these H-bonds are either not formed yet or so unstable that the amides become protonated during the labeling pulse. However, we cannot exclude the presence of other types of interactions, such as hydrophobic side chain contacts or interactions involving the heme group.

It seems surprising that association of two helical segments near opposite ends of the chain is preferred over other more local contacts, especially since in the native cyt c structure (21) the contact area between the 60's helix and the C helix is comparable to that between the N-helix and the C-helix. However, the pairing site between the N-helix and the C-helix appears to be tighter than that between the 60's helix and C-helix. The closest approach occurs near Gly 6 and Leu 94. Interestingly, Gly 6 is one of the few completely conserved residues among all known c-type cytochromes, and the hydrophobic character of the other contact residues is also highly conserved. The helix pairing site is a prominent feature of all known cyt c structures, even of the very distant prokaryotic relatives. Thus, it is tempting to speculate that the conservation of this structural feature reflects constraints imposed by the folding mechanism.

Another possible explanation for the late formation of the 60's helix is related to the involvement of heme ligation in cyt c folding. A number of observations (Elöve & Roder, to be published) indicate that Met 80 which is ligated to the heme iron in the native cyt c structure is replaced by another side chain ligand, probably His 26 or His 33, in the unfolded state. This non-native axial ligand becomes trapped in the partly folded early intermediate, and its presence may interfere with the proper folding of the 60's region and other parts of the structure. The non-native ligand is replaced by the native methionine in the intermediate phase on the 100 ms time scale (34; Elöve & Roder, in preparation).

It is interesting to note that the fluorescence of Trp 59 is partially quenched in the early folding phase (Figure 3), while the indole NH of the same tryptophan only begins to be protected in the intermediate phase by forming an H-bond with one of the heme propionate side chains (Figure 2). Apparently, the early intermediate formed in 20 ms is sufficiently compact for the fluorescence to become partially quenched via energy transfer with the heme, but a stable H-bond is not yet present.

Variation of the Pulse Labeling Conditions

The preferential protection of N- and C-terminal amide sites in the fast kinetic phase is a clear indication of a sequential folding reaction with a partly folded intermediate on the folding pathway. On the other hand, the pulse labeling results also show indications of heterogeneous behavior with several species refolding along parallel pathways. For instance, N- and C-terminal sites acquire about 60% protection in the fast phase, suggesting that 60% of the molecules have undergone the helix pairing reaction while the remaining 40% are still unfolded after 30 ms. Similarly, the 50% occupancy observed for amide protons in the 60's helix at t_f=500 ms is consistent with a heterogeneous population in which about half of the molecules are completely folded while 30-40% of the molecules are structured only at the chain termini and 10-20% are completely unfolded (Figure 2). However, we have to consider another possibility, namely that the system is homogeneous, but amide sites are only partially protected. For example, a proton occupancy of 0.5 for a particular NH could arise if the degree of protection is such that the amide becomes 50% protonated during the labeling pulse. These two possibilities can be distinguished by a modified pulse labeling protocol involving systematic variation of the labeling conditions. Furthermore, this approach can provide qualitative information on the stability of folding intermediates. A similar method has been applied to ribonuclease A (9).

The analysis of pulse variation experiments is based on a structural unfolding model which has been used extensively in the interpretation of H-exchange data from native proteins (e.g. *13, 15-16*). While some of the details of the hydrogen exchange mechanism are still unclear, there is general agreement that under destabilizing conditions where the equilibrium between folded molecules (F) and unfolded forms (U) is well approximated by a two-state model, the strongly protected protons exchange via transient unfolding (4). This situation can be described by the following scheme:

$$F(H) \underset{}{\overset{K_u}{\rightleftharpoons}} U(H) \underset{D_2O}{\overset{k_c}{\longrightarrow}} U(D) \rightleftharpoons F(D) \tag{1}$$

where K_u represents the equilibrium constant for unfolding (disruption of intramolecular H-bonds) and k_c is the exchange rate of a given amide site in the unfolded state. Under most conditions, k_c is rate limiting (EX_2 exchange mechanism), so that the measured exchange rate, k_{ex}, is given by

$$k_{ex} = K_u k_c. \tag{2}$$

NH exchange rates can thus provide a measure of the free energy of unfolding, ΔG_u, according to the relationship

$$\Delta G_u = -RT \ln(k_{ex}/k_c). \tag{3}$$

The validity of Equation (3) has been demonstrated in a study of the effect of destabilizing solvent additives on NH exchange rates in native BPTI (4), and it is expected to be applicable to transient folding intermediates as well.

NH exchange rates in folding intermediates can in principle be measured by pulse labeling experiments at constant folding time and variable pulse time (t_p), but due to the limited life time of a folding intermediate it is advantageous to vary pulse pH at constant t_p. For a homogeneous situation, the resulting pH dependent proton occupancy, $P(pH)$, is given by

$$P(pH) = 1 - \exp[-K_u k_c(pH) t_p]. \tag{4}$$

Above pH 3, the exchange reaction is determined by base catalysis (*11-13*), and k_c can be expressed as

$$k_c(pH) = k_{OH} 10^{(pH-pK_w)}, \tag{5}$$

where k_{OH} is the second order rate constant for base catalyzed exchange. In the absence of pH dependent conformational transitions (constant K_u), Equation (4) describes a sigmoidal curve with $P=0$ at low pH and $P=1$ at high pH, as shown in Figure 5 (left panel). For an exposed amide proton ($K_u=1$) under the conditions used in this study ($t_p=30$ ms, $10°C$), the labeling profile is expected to be centered near pH 8 (where the half life for exchange matches t_p). With increasing stability ($K_u \ll 1$), the curve shifts towards higher pH. In the case of a heterogeneous mixture of molecules with different degrees of protection (Figure 5, right panel), one expects multiphasic pH profiles for a given amide proton, i.e. $P(pH)$ undergoes several stepwise increments with increasing pulse pH.

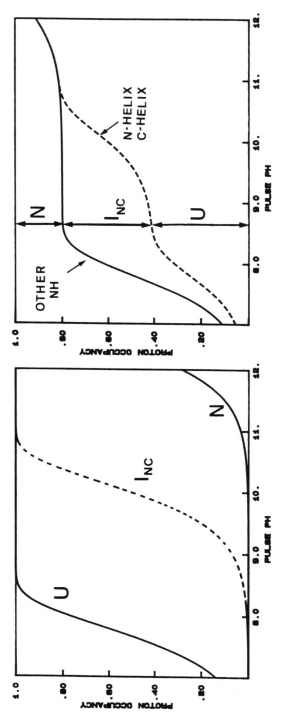

Figure 5. Model calculations based on Equation 5, of the pulse-pH profiles expected for a homogeneous situation (left panel) and a heterogeneous mixture of 20% native-like molecules, 40% partially folded intermediates and 40% unfolded molecules (right panel).

Pulse pH variation studies were carried out at a folding time of 30 ms where the early intermediate is well populated (c.f. Figure 2). Figure 6 shows the proton occupancy as a function of the pulse pH for several NH resonances that can be resolved by 1D ^{1}H NMR, including representative amide protons from the three major α-helices in cyt c. Below pH 7.5, the occupancies are low for all protons since the rate is too slow for any D-H exchange to occur during the 30 ms labeling time. At pH 8.5, the exchange time constant of the intrinsic exchange reaction is about 10 ms so that exposed amide sites become protonated. Above pH 8.5, a nearly pH-independent plateau sets in with an occupancy of about 0.4 for class 1 protons and 0.8 for class 2 protons. This behavior is clearly inconsistent with a single sequential folding pathway, but indicates the presence of a heterogeneous mixture of unfolded molecules (40%), partially folded species (40%) and fully folded molecules (20%). The increase in proton occupancies above pH 10 indicates that protected amide protons are beginning to become labeled as the intrinsic exchange rates increase further , approaching microseconds at pH 12. The pulse profiles at basic pH provide information on the degree of protection, and thus the stability of the structure in the various species of the folding mixture. For example, the results for Val 11 and Leu 98 are consistent with a stability of about 3 kcal/mol for the N- and C-terminal helices in the early folding intermediate. Moreover, identical pH profiles are observed for protons on either helix, indicating that they are protected within the same structural unit. The fluctuations that lead to exchange, apparently involve transient dissociation of the preformed complex between the N-helix and the C-

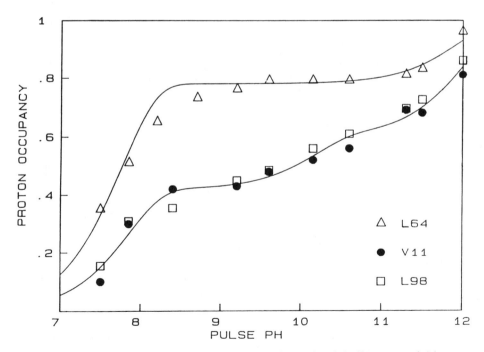

Figure 6.   Pulse-pH profiles measured by pulse labeling at variable labeling pH after a refolding time of 30 ms. Proton occupancies measured by 1D NMR are shown for representative amide sites in the three main α-helices of cyt c.

helix. Together with the observation of identical protection kinetics for N- and C-terminal protons (Figure 2), this strongly supports the presence of a helix pairing site in the early folding intermediate of cyt c. It would be highly unlikely for two non-interacting helices to form simultaneously and to exhibit the same apparent stability. These results underscore the importance of tertiary interactions in stabilizing elements of secondary structure, even at an early stage in folding.

The NH occupancy for Leu 64 (Fig. 6) as well as other protons in the 60's helix and some protons involved in non-helical H-bonds (not shown) levels off at a value of 0.8 above pH 9. This level increases only above pH 11.5 where labeling is expected to occur even in the native state. Apparently, within 30 ms of refolding, about 20% of the molecules have already acquired extensive H-bonded structure with a stability similar to that of the native state. The kinetic labeling curves in Figure 2 show that most NH sites are already about 20% protected at the first folding time measured (~ 3 ms), suggesting that the rapidly folding population is formed within the mixing dead-time.

Acknowledgments

This work was supported by a grant from the National Institutes of Health (GM35926). We thank Paul B. Laub for discussions and molecular graphics calculations.

Literature Cited

1. Kim, P.S.; Baldwin R. L. *Ann. Rev. Biochem.* **1990**, *59*, 631-650.
2. Alber, T. *Ann. Rev. Biochem.* **1989**, *58*, 765-798.
3. Goldenberg, D. *Ann. Rev. Biophys. Biophys. Chem.* **1988**, *17*, 481-507.
3. Kuwajima, K. *Proteins* **1989**, *6*, 87-103.
4. Roder, H. *Meth. Enzymol.* **1989**, *176*, 446-472.
5. Roder, H. and Wüthrich, K. **1986**, *Proteins 1*, 34-42.
6. Schmid, F. X.; Baldwin R. L. *J. Mol. Biol.* **1979**, *135*, 199-215.
7. Roder, H.; Elöve, G. A.; Englander, S. W. *Nature* **1988**, *335*, 700-704.
8. Udgaonkar, J. B.; Baldwin, R. L. *Nature* **1988**, *335*, 694-699.
9. Udgaonkar, J. B.; Baldwin, R. L. *Proc. Natl. Acad. Sci. USA*, **1990**, *87*, 8197-8201.
10. Bycroft, M.; Matouschek, A.; Kellis, J. T. Jr; Serrano, L.; Fersht, A. R. *Nature* **1990**, *346*, 488-490.
11. Molday, R. S.; Englander, S. W.; Kallen, R. G. *Biochemistry* **1972**, *11*, 150-158.
12. Roder, H.; Wagner, G.; Wüthrich, K. *Biochemistry* **1985**, *24*, 7407-7411.
13. Englander, S. W.; Kallenbach, N. R. *Q. Rev. Biophys.* **1984**, *16*, 521-655.
14. Kim, P. S. *Meth. Enzymol.* **1986**, *31*, 136-156.
15. Wagner, G.; Wüthrich, K. *J. Mol. Biol.* **1982**, *160*, 343-361.
16. Wand, A. J.; Roder, H.; Englander, S. W. *Biochemistry* **1986**, *25*, 1107-1114.
17. Jeng, M-F.; Englander, S. W.; Elöve, G. A.; Wand, A. J.; Roder, H. *Biochemistry* **1990**, *29*, 10433-10437.
18. Swanson, R.; Trus, B. L.; Mandel, N.; Mandel, G.; Kallai, O. B.; Dickerson, R. E. *J. Biol. Chem.* **1977**, *252*, 759-775.
19. Ochi, H.; Hata, Y.; Tanaka, N.; Kakudo, M.; Sakuri, T.; Aihara, S.; Morita Y. *J. Mol. Biol.* **1983**, *166*, 407-418.
20. Louie, G. V.; Hutcheon, W.; Brayer, G. D. *J. Mol. Biol.* **1988**, *199*, 295-314.
21. Bushnell, G.W.; Louie, G. V.; Brayer, G. D. *J. Mol. Biol.* **1990**, *214*, 585-595.
22. Babul, J.; Stellwagen, E. *Biopolymers* **1971**, *10*, 2359-2361.
23. Babul, J.; Stellwagen, E. *Biochemistry* **1972**, *11*, 1195-1200.

24. Ikai, A.; Tanford, C. *J. Mol. Biol.* **1973**, *73*, 145-164.
25. Ikai, A.; Fish, W. W.; Tanford, C. *J. Mol. Biol.* **1973**, *73*, 165-184.
26. Fisher, W. R.; Tanuichi, H.; Anfinsen, C. B. *J. Biol. Chem.* **1973**, *248*, 3188-3195.
27. Tsong, T. Y. *Biochemistry* **1976**, *15*, 5467-5473.
28. Stellwagen, E. *J. Mol. Biol.* **1979**, *135*, 217-229.
29. Henkens, R. W.; Turner, S. R. *J. Mol. Biol.* **1979**, *254*, 8110-8112.
30. Ridge, J. A.; Baldwin, R. L.; Labhardt, A. M. *Biochemistry* **1981**, *20*, 1622-1630.
31. Brems, D. N.; Stellwagen, E. *J. Biol. Chem.* **1983**, *258*, 3655-3660.
32. Wand, A. J.; DiStefano, D. L.; Feng, Y.; Roder, H.; Englander, S. W. *Biochemistry* **1989**, *28*, 186-194.
33. Feng, Y.; Roder, H.; Englander, S. W.; Wand, A. J.; DiStefano, D. L. *Biochemistry* **1989**, *28*, 195-203.
34. Brems, D. N.; Stellwagen, E. *J. Biol. Chem.* **1983**, *258*, 3655-3660.
35. Shoemaker, K. R.; Kim, P. S.; York, E. J.; Stewart, E. J.; Baldwin, R. L. *Nature* **1987**, *326*, 563-567.
36. Wright, P. E.; Dyson, H. J.; Lerner, R. A. *Biochemistry* **1988**, *27*, 7167-7175.

RECEIVED March 11, 1991

Chapter 5

Thermostability of Drifted Oligomeric, Reduced, and Refolded Proteins

S. L. Sagar and M. M. Domach

Department of Chemical Engineering, Carnegie Mellon University, Pittsburgh, PA 15213

Renatured preparations of monomeric, crosslinked proteins can be heterogeneous. Processes such as conformational drift have also been suggested to occur during the purification and storage of oligomeric proteins. Drift results in heterogeneity and a loss of enzymatic activity, but in some cases renaturation is possible. To complement existing characterization methods, ligand binding and DSC analysis was used to probe for the presence of species that differed in thermal stability. Yeast and liver alcohol dehydrogenases were examined as well as native, reduced, and refolded lysozyme. Multiple species were found in the yeast enzyme samples and lysozyme refolded under nonoptimal conditions. The liver enzyme exhibits two thermal transitions which is consistent with the presence of drifted dimers, but at this time the assignment of the origin of the transitions is tentative.

Protein renaturation is a multifaceted problem. Minimizing the extent of aggregation during refolding and contending with contaminating proteases that coprecipitate in the intracellular inclusion body with the desired protein are among the technical challenges addressed by this symposium. In addition to these problems, information on the degree of homogeneity before and after each renaturation step is required so that the activity yields can be interpreted and the design of a renaturation process can be optimized in order to minimize heterogeneity. Additionally, extending renaturation methods, if possible, to oligomeric proteins will require knowledge about the dynamics of oligomeric proteins and the means of characterizing the constituents of the population.

Inhomogeneity can arise from many sources. For example, even if the protein is not produced by the cell in an inclusion body, thereby obviating the need for resolubilization and refolding, translational errors can occur. These errors will result in the synthesis of proteins that differ in their primary amino acid sequence. The potential outcome is that not all protein molecules will have the same activity and/or stability. The extent to which differences will be apparent will depend on what residues have been altered, their location (e.g., core, surface, or active site), the protein's tolerance to substitutions (e.g., (1)), and the detection method used (e.g. activity assay vs. fractionation). For proteins produced as inclusion bodies, nonoptimal refolding conditions can result in the production of small aggregates or incorrect crosslinking patterns. An interesting example of how the crosslinking pattern of a refolded enzyme can vary is included in this symposium.

0097–6156/91/0470–0064$06.00/0

Regarding oligomeric proteins, heterogeneity can be the outcome of alternate processes that occur during purification and storage. One denaturation mechanism is the "conformational drift process" (*2,3*). The process is believed to entail continual oligomer dissociation and reassociation with the result that "conformationally drifted species" eventually accumulate. The drifted species are free subunits with nonnative conformations and oligomers containing drifted subunits. The results of pressure-induced denaturation experiments provide the evidence for conformational drift. A pressure ramp is applied which results in dissociation at low pressure and ultimately unfolding of free subunits at high pressure. By observing the fluorescent emission of the sample as a function of pressure, the tendency of the protein molecules in a sample to dissociate and unfold can be determined. Thus, the stability spectrum of the species present and how the spectrum changes after a sample has experienced different histories can be established. For example, storage of lactate dehydrogenase (LDH) at 4°C results in a reduction of enzymatic activity (*2*). LDH tetramers were also found to dissociate at pressures lower than that of solutions with high activity. Additionally, the results of gel filtration experiments indicated that the tetramers and monomers present are likely active tetramers and drifted subunits as opposed to a slowly equilibrating mixture of active tetramers and dissociated subunits. Interestingly, it was also shown that through the use of temperature-time treatment, nearly full restoration of LDH enzymatic activity could be accomplished. Conformational drift of glyceraldehydephosphate dehydrogenase (GAPDH) has also been observed (*3*). The measured variance in subunit binding energy was significant and indicative of the extent of drift.

Because heterogeneity in the renatured product is a practical concern, fractionation methods, antibody binding, and activity assays are frequently employed to characterize a protein preparation. However, fractionation methods that rely on differences in Stokes-Einstein radii for the resolution of different molecular species can be insensitive. Instead, affinity or hydrophobic chromatography may be required to resolve the species present. To attempt to complement these methods, we have employed differential scanning calorimetry (DSC) for characterizing the heterogeneity of protein samples. In practice, a sample and reference are subjected to the same temperature-time ramp. The unfolding of a protein's domain(s) is an endothermic process and the unfolding event(s) are detected by an increase in sample heat capacity. The temperature at which the excess heat capacity is maximal (T_m) is indicative of the protein's stability and the heat capacity curve can provide information on the number of underlying cooperative unfolding processes (for review, *see* (*4*)). It was thus envisioned that DSC could prove useful because protein solutes are distinguished by their denaturation temperatures which, in turn, depend on molecular conformation.

To examine the utility of DSC, experiments with oligomeric and monomeric proteins were conducted. The results of the DSC analysis of tetrameric yeast alcohol dehydrogenase (YADH) are reported first. YADH is oligomeric and a dehydrogenase like LDH and GAPDH; hence, the conformational drift process may also take place. It was envisioned that the resolution and identification of YADH species with differing thermal stabilities would be aided by using coenzyme as a probe. That is, because coenzyme will preferentially bind to native-like coenzyme binding domains, and ligand binding tends to alter a protein's denaturation temperature (e.g., (*4,5*)), improved resolution of the different thermal transitions should result when coenzyme is present. It should be noted, however, that any improvement in resolution is not necessarily the result of the thermal stability of the ligand binding species (or domain) increasing due to the binding of a ligand. Rather, if a ligand binds preferentially to a native conformation, the van't Hoff equation predicts that the temperature at which half the native species will be unfolded will increase (*5*). Partial scan-rescan experiments and scanning with 1M urea present were also conducted to improve resolution as well as provide information on the origin of the transitions. Then, the results of some preliminary DSC experiments on another dehydrogenase, dimeric liver alcohol dehydrogenase (LADH), are described.

Lysozyme was the subject of the experiments with a monomeric protein. Prior DSC work has shown that lysozyme can undergo a reversible, two-state transition over the pH 2-5 range and that the denaturation temperature increases as pH is increased (6,7). Prior DSC work also has utilized reduced lysozyme in order to minimize the constraints the crosslinks offer to random coil formation. In contrast, we characterized the reduced form by again using ligand binding as a probe. Determining the extent to which a reduced enzyme can retain native ligand-binding properties was the desired outcome. Additionally, it was of interest to determine how the scans would reflect nonoptimal refolding conditions. To accomplish this, lysozyme was annealed at high temperature after being unfolded and pH was fixed at the alkaline side of the range where refolding is essentially complete. After this treatment, refolding at room temperature was allowed to occur. By using these conditions, events such as proline isomerization (lysozyme contains two proline residues) and disulfide exchange (native enzyme has four crosslinks), but not aggregation, were anticipated to lead to incomplete refolding (8,9) and an increased possibility that multiple forms would be apparent in the scans of refolded enzyme.

Experimental

Yeast alcohol dehydrogenase (product number A-7011), lysozyme (product number L7001), dithiothreitol (DTT), N-acetylglucosamine (GlcNAc), and oxidized coenzyme (NAD+) were obtained from Sigma Chemical Company (St. Louis, MO.) and used without further purification. Liver alcohol dehydrogenase (product number 11305427) was obtained from Boehringer. Prior to scanning, an enzyme was first dialysed overnight against the buffer. The enzyme was then filtered through a 0.2 μM pore size membrane to remove any insoluble material. Enzyme samples and references were degassed prior to injection into the calorimeter by the use of an aspirator.

A MicroCal MC-2 scanning calorimeter (MicroCal Inc., North Hampton, MA.) was used to perform the thermal denaturation experiments. Sample and reference volume used was 1.2 ml and enzyme concentrations ranged from 1 to 5 mg/ml. Baseline and shift settings were adjusted to give a flat water-water baseline. Cell feedback was fixed at a value of 32 and 16X sensitivity and 0.5°C/min scan rate were used. Following a scan, a 4X calibration pulse was used to scale the endotherm; in practice, little instrumental drift was observed.

A 30 cm Protein Pak 300 SW column (Waters product no. 80013) and isocratic Waters HPLC were used to perform the chromatography studies. Solvent flow rate was fixed at 1.0 ml/min and 10 ml samples of 1-10 mg/ml were injected. Eluted components were detected by uv absorption at 280 nm. The retention times of myoglobin (18,800 MW), α-chymotrypsin (25,000 MW), hexose kinase (100,000 MW), and glucose oxidase (186,000 MW) were determined in order to calibrate the column.

Oligomeric Protein Results

YADH Scans in the Absence and Presence of Oxidized Coenzyme. The pH range 6.26 to 9.4 was examined. The scans obtained when coenzyme was absent or present at a concentration of 10^{-3} M are shown in Figure 1. The scans reveal that at least two transitions occur except at pH 9.4 where they are not resolved. For the pH range examined, the way in which the T_m of the dominant transition depends on pH and the presence of NAD+ is summarized in Figure 2. As anticipated, the addition of coenzyme results in increasing the temperature separation of the transitions except at the extreme ends of the pH range investigated. Additionally, because YADH is most active at pH 8.8, the effect of NAD^+ increasing T_m when the pH is in the vicinity of the activity maxima suggests that the YADH species or domains that can bind NAD^+ are responsible for the high T_m transition.

The observation of at least two transitions is likely not due to the presence of a protein contaminant. Our gel electrophoresis analysis of sodium dodecyl sulfate-treated YADH yielded only one band that corresponded to the molecular weight of a YADH subunit (10). Thus, the presence of labile species or domain(s) in the sample must account for the two transitions. For example, the low T_m transition could be attributed to the denaturation of a labile dissociated subunit or a labile domain in the tetramer. Such labile subunits or domains may be akin to the conformationally drifted species found for other dehydrogenases (2,3). The high T_m transition, in turn, may be attributable to the denaturation of an active tetramer, or coenzyme binding domain, because the transition exhibited the greatest response to NAD^+. Other alternatives are that tetrameric YADH reversibly dissociates and the subunits are responsible for the low T_m transition, or that the low and high T_m transitions can be attributed to thermally-induced dissociation and denaturation of the subunits, respectively. To aid the interpretation of the scans, additional DSC experiments were performed.

YADH Partial Denaturation Experiments. When the sample was partially scanned (i.e., increase the temperature from 16^oC to 57^oC and then cool down to 16^oC) and then rescanned over the full temperature range after a period of 5 hours, only the high T_m transition was observed (Figure 3). If reversible dissociation occurred to a significant extent, and the free subunit was more labile than the tetramer, then one would expect that the low T_m transition would ultimately reappear in the second scan due to the dissociation of the tetramer replacing the irreversibly denatured subunits. Likewise, the low T_m transition can be eliminated by treating the sample with 1 M urea prior to scanning (Figure 4). The results of these experiments thus show that the low T_m transition can not be restored by annealing the sample after exposure to mild denaturing conditions. Finally, the relative magnitudes of the two transitions were insensitive to YADH concentration over the range, 1 - 10 mg/ml (not shown). If reversible dissociation was a factor, then increasing the enzyme concentration should lower the extent of monomer formation. Taken together, the results of additional DSC experiments suggest that denaturation of reversibly dissociated subunits can not account for the low T_m transition. However, the possibility that a labile domain in a native or drifted tetramer is present can not be eliminated by the DSC results. Therefore, size exclusion chromatography experiments were also performed.

YADH Size Exclusion Chromatography Experiments. The chromatography results indicate that a high retention time component with a molecular weight corresponding to a YADH subunit is present (Figure 5). Interestingly, the chromatogram is similar to one reported for LDH (2) which has been found to contain conformationally drifted, dissociated subunits. The relative abundance of the two components in the chromatograms was not significantly altered when 1 mg/ml or 10 mg/ml samples were injected (not shown). If reversible dissociation was a significant factor, then a ten-fold dilution of the sample should have had an effect on the relative abundance of the two components eluted. Moreover, dilution during elution from a fractionation column can be substantial; hence, if dissociation was a significant factor, then it would be more apparent in the chromatography experiments than the scan-rescan DSC experiments. Finally, our size exclusion chromatography studies also indicate that a relationship between the relative abundance of the two components and the age and extent of hydration (i.e. effectiveness of desiccation during storage) of the crystalline preparation appears to exist (not shown). The chromatography results thus indicate that essentially irreversibly dissociated subunits are present in the YADH

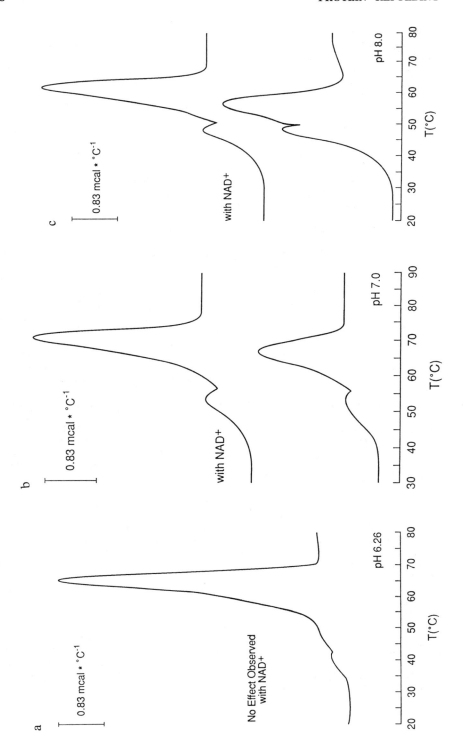

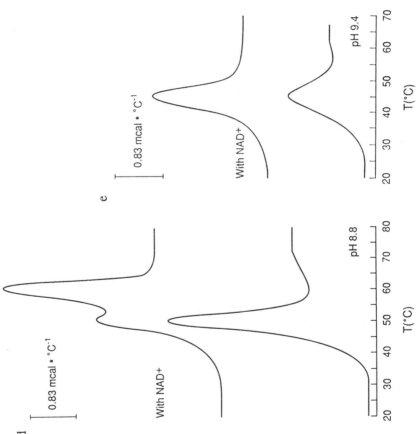

Figure 1. Scans of yeast alcohol dehydrogenase at various values of pH and in the presence or absence of 10^{-3} M NAD$^+$. Values of pH are (a) 6.26, (b) 7.0, (c) 8.0, (d) 8.8, and (e) 9.4.

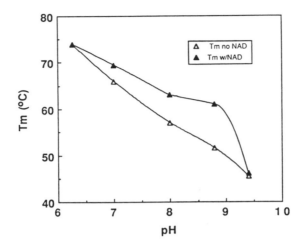

Figure 2. Variation of denaturation temperature of the major endotherm (T_m) with pH when 10^{-3} M NAD$^+$ is present or absent. NAD$^+$ increases the denaturation temperature while increasing pH tends to decrease the denaturation temperature.

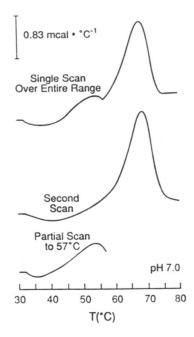

Figure 3. Partial scan-rescan experiment on yeast alcohol dehydrogenase at pH 7. Scanning to 57°C results in one transition being observed (bottom). Rescanning the same sample results in the absence of low temperature transition (middle) while the high temperature transition is largely unchanged compared to the scan over the full temperature range (top).

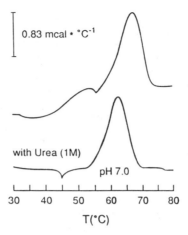

Figure 4. Effect of 1 M urea on endotherm of yeast alcohol dehydrogenase. Addition of urea results in the elimination of the low temperature transition (bottom) as compared to the scan in the absence of urea (top). A low temperature exotherm is also observed which can be attributed to the aggregation of the thermally denatured subunit.

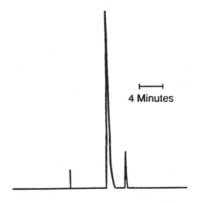

Figure 5. Size exclusion chromatogram of yeast alcohol dehydrogenase. The time at which the sample was injected is denoted by the vertical line. Two components elute where one has a retention time between that of α-chymotrypsin (25,000 MW) and liver alcohol dehydrogenase (80,000 MW) and the other's time is between that of liver alcohol dehydrogenase and glucose oxidase (200,000 MW).

sample which suggests that the low T_m transition can be plausibly attributed to their denaturation.

LADH Thermal Denaturation. The endotherms of LADH at various pH values are shown in Figure 6. As was observed for YADH (*see* Figure 1), two transitions are apparent in several of the LADH endotherms. Additionally, the temperature at which unfolding occurs decreases as the pH is increased. However, unlike YADH, only one species that had a molecular weight corresponding to the LADH dimer was found to elute during size exclusion chromatography (not shown). The fact that no subunits were observed does not preclude the possibility that different drifted dimers are present in the LADH sample. Rather, it has been noted by Weber and colleagues that the extent of dissociation in some cases is low; hence, only after substantial drift has occurred will drifted subunits be present (2). Thus, it is possible that at least two dimeric species of LADH that differ in thermal stability are present in the sample. This hypothesis is currently under investigation.

Lysozyme Results

Reduced Lysozyme Thermal Denaturation. A scan of native lysozyme is shown in Figure 7a. The T_m is 82.5 $^\circ$C and the enthalpy ratio (the ratio of the calorimetric enthalpy to the apparent van't Hoff enthalpy) is approximately one which agrees with prior DSC analyses of lysozyme (6,7). Additionally, a significant change in heat capacity occurs following unfolding which is characteristic of lysozyme (4). When lysozyme's disulfide bonds are reduced by the addition of DTT, a 7°C decrease in T_m occurs (Figure 7b), but the enthalpy ratio is not significantly different from that of the crosslinked enzyme. The decrease in T_m is to be expected because stabilizing crosslinks have been removed. Overall, in the presence of DTT a significant amount of secondary and tertiary structure is retained by lysozyme and the molecular population is quite homogeneous as reflected by the observation of only one transition and a constant enthalpy ratio.

In order to investigate reduced lysozyme further, the response of the reduced enzyme to the ligand N-acetylglucosamine (GlcNAc) was compared to that of the native enzyme. As seen for YADH, ligands that bind favorably to native proteins often cause an increase in T_m and/or denaturation enthalpy to occur. GlcNAc binding to reduced lysozyme does indeed result in a T_m and enthalpy change (Figure 8b) as it did for the native enzyme (Figure 8a). Thus, it appears that the reduced enzyme retains the native enzyme's ability to bind a ligand.

Refolded Lysozyme Thermal Denaturation. When lysozyme was thermally denatured at pH 5.68 and annealed at 100°C, precipitation was not observed, but incomplete refolding occurred. For example, a scan of refolded lysozyme after thermal denaturation is shown in Figure 9. In comparison to the native enzyme (*see* Figure 7a), the endotherm is not as sharp and a low temperature shoulder is also evident. We attribute the broadness of the endotherm of the refolded enzyme to disulfide exchange and/or incorrect secondary structure formation.

Discussion

We used DSC and ligand binding to probe the heterogeneity and thermal stability of oligomeric, reduced globular, and refolded globular proteins. Multiple thermal transitions were detected for both YADH and LADH. In the case of YADH, the transitions could be plausibly attributed to the presence of drifted subunits and tetramers. The assignment was based on the results of size exclusion chromatography, coenzyme addition, scan-rescan, and urea addition experiments. Additionally, it was

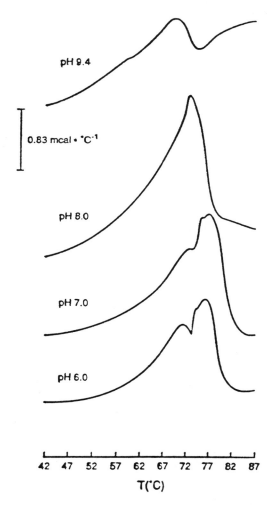

Figure 6. Scans of liver alcohol dehydrogenase at various values of pH.

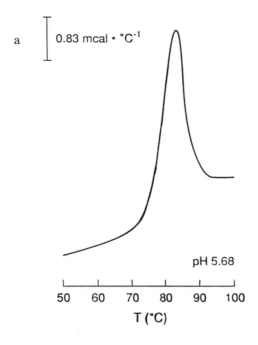

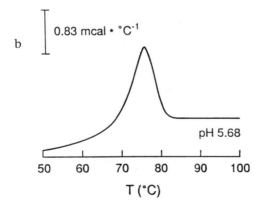

Figure 7. Endotherms of (a) native and (b) reduced lysozyme. In the presence of dithiothreitol, the denaturation temperature decreases.

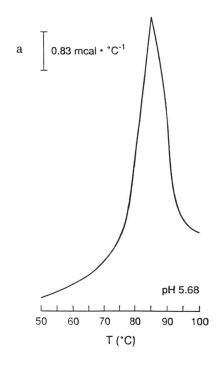

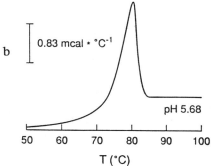

Figure 8. Effect of the ligand, N-acetylglucosamine, on the thermal unfolding of (a) native and (b) reduced lysozyme. In comparison to the case where the ligand was absent (*see* Figure 7), an increase in denaturation temperature occurs for the native (3^oC) and reduced (5^oC) enzymes when the ligand is present. The increase indicates that the reduced enzyme is able to bind the ligand.

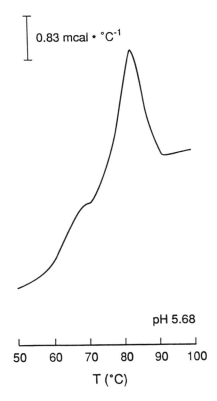

Figure 9.   Scan of refolded lysozyme which was denatured, annealed, and allowed to refold at pH 5.68. In comparison to the native enzyme (*see* Figure 7a) the endotherm is less monodisperse and appears to contain a low temperature shoulder.

noted that the results of chromatography experiments indicate that the abundance of "drifted" YADH subunits increases with time. For LADH, the multiple transitions were tentatively attributed to the presence of different dimeric species. Work is in progress to evaluate the LADH assignment further. However, the DSC results are consistent with the results of high pressure-fluorescence spectroscopy experiments (*2,3*). Regarding the characterization of YADH, the DSC and chromatography analyses complemented each other.

Although only one globular protein has been examined to date, some interesting results have been obtained. Although reduced lysozyme has less thermal stability than the native enzyme, it still exhibits ligand binding ability. Additionally, as others have found, the reduced enzyme is quite homogeneous based on its enthalpy ratio being equal to the native enzyme's and only one symmetric transition being observed. In contrast, the scan of lysozyme that was refolded under nonoptimal conditions exibits a shoulder suggesting that multiple forms are present. These multiple forms likely have different crosslinking patterns.

Extending the implications of the lysozyme experiments to the recovery of crosslinked proteins from inclusion bodies is tenuous because the reduced intermediate formed during the recovery process may experience a different history than a native enzyme that has had its accessible disulfides reduced. However, if the primary amino acid sequence determines a protein's structure and crosslinking is a "finishing step", then a deviation of the endotherm of a reduced intermediate in recovery process from that of a reduced native enzyme could be an early indication that the refolding conditions will not result in correct precursor formation. Additionally, a change in enthalpy ratio or scan modality following the oxidation of a reduced intermediate could be indicative of a nonuniform crosslinking pattern being generated in the renaturation steps. In this case, DSC analysis could also complement the results of hydrophobic chromatography or PAGE experiments. For example, the protein species that is retained the most on a hydrophobic column could be hypothesized to be the one most able to unfold and expose hydrophobic residues due to the number of crosslinks (or pattern) not yielding maximal stability. DSC analysis would corroborate the hypothesis if the T_m of the isolate with the longest retention time proves to be the smallest. Moreover, the DSC results could be used to help identify the nature of each species retained on the chromatography column. To further pursue these implications, we are currently examining the enzyme, TEM-β-lactamase, which is expressed at a high level by recombinant *E. coli*. PAGE analysis of samples from either resolubilized inclusion bodies or cell extracts indicates that the enzyme is recovered two forms (*11*). The molecular weights are similar; hence, the forms differ in their crosslinking pattern. Subsequent fractionation and DSC experiments may yield information on the number of crosslinking patterns present in the sample.

Acknowledgments

This work was initiated by NSF Grant CPE-8513443 and continued support provided, in part, by NSF Grant CBT-8657533.

Literature Cited

1. Bowie, J.U.; Reidhaar-Olson; J.F., Lim, W.A.; & Sauer, R.T. *Science* **1990**, *247*, 1306.
2. King, L. & Weber, G. *Biochemistry* **1986**, *25*, 3637.
3. Ruan, K. & Weber, G. *Biochemistry* **1989**, *28*, 2144.
4. Sturtevant, J.M. *Ann. Rev. Phys. Chem.* **1987**, *38*, 463.
5. Edge, V.; Allewell, N.M.; & Sturtevant, J.M. *Biochem.* **1985**, *24*, 5899.

6. Privalov, P.L. *Advances in Protein Chemistry* **1979**, *33*, 167.
7. Privalov, P.L. & Khechinashvili, N.N., *J. Mol. Biol.* **1974**, *86*, 665.
8. Jainicke, R. *Prog. in Biophys. and Mol. Biol.* **1987**, *49*, 117.
9. Creighton, T.E. *Methods in Enzymol.* **1986**, *131*, 83
10. Kennedy, C.L.; Sagar, S.L.; Zanapolidou, R.A.; Tristram-Nagle, S.; & Domach, M.M. *Biotechnol. Prog.* **1989**, *5*, 164.
11. Personal communication, M.A. Ataai, Univ. of Pittsburgh.

RECEIVED February 6, 1991

Chapter 6

Oxidative Refolding of Recombinant Human Glia Maturation Factor Beta

Asgar Zaheer and Ramon Lim

Division of Neurochemistry and Neurobiology, Department of Neurology,
University of Iowa College of Medicine, Iowa City, IA 52242

Recombinant human glia maturation factor beta (r-hGMF-beta) expressed in *E. coli* is a single-chain polypeptide containing three cysteine residues (at amino acid positions 7, 86 and 95) out of a total of 141 amino acids. It possesses no disulfide bonds nor biological activity. After oxidative refolding in the presence of a " redox " system consisting of reduced and oxidized forms of glutathione, in the presence of guanidine hydrochloride, biological activity appears. Reverse-phase HPLC analysis of the refolded r-hGMF-beta shows the presence of four peaks. We speculate that these peaks correspond to the reduced form plus three isoforms containing intramolecular disulfide bonds as predicted from the number of cysteine residues. Only one of the three disulfide-linked isoforms exhibits biological activity as determined by proliferation inhibition assay on C6 glioma cells.

In recent years the concept of biotherapy, that is, use of biologicals produced by mammalian cells to obtain a favorable response in the target cells, has generated tremendous interest in medicine. These biologicals include a variety of agents such as lymphokines, cytokines, tumor associated antigens, tumor suppressive agents, and growth and maturation factors. With the development of new technologies, including recombinant DNA techniques, a large number of agents such as interferons, interleukins, colony stimulating factors, tumor necrosis factor, fibroblast growth factor, transforming growth factor, platelet-derived growth factor and nerve growth factor, are now available in large quantities for laboratory and clinical trails. The biological activity of proteins produced by recombinant DNA method is not only dependent on correct amino acid sequence but also on efficient and correct folding of the polypeptide chain to form proper tertiary structure. One of the problems encountered in recombinant proteins is substantial reduction or sometimes total loss of biological and/or immunological activity as compare to their natural counterpart. In case of proteins known to contain disulfide bonds essential for bioactivity, the problem seems to originate from either no disulfide bond or incorrectly formed disulfide bonds. For example, formation

0097–6156/91/0470–0079$06.00/0

of correct intramolecular disulfide bond is necessary to obtain the biologically active forms of interleukin-2 (1), interleukin-6 (2), granulocyte colony stimulating factor (2), insulin-like growth factor II (3) and interferons (4,5).

Glia maturation factor beta (GMF-beta), first detected in this laboratory in 1972 (6-8), is an acidic protein with a molecular mass of 17 kDa present in brains of all vertebrates examined (9,10). GMF-beta promotes the differentiation of glia and neurons, and inhibits the growth of their corresponding tumors (11,12). Although GMF-beta was purified to homogeneity from bovine brain (11), it was not available in sufficient quantity for structural studies. Recently, GMF-beta has been completely sequenced (13), cloned and expressed in E. coli (14,15). Pure recombinant human GMF-beta (r-hGMF-beta) is in fully reduced form with no biological activity. In the present paper we describe a method for oxidative refolding of r-hGMF-beta, with generation of three disulfide-bonded isomeric forms, and the isolation of the biologically active isoform by reverse-phase HPLC.

The recombinant human glia maturation factor beta (r-hGMF-beta) expressed in E. Coli has identical amino acid composition and sequence with the natural bovine brain protein (13). The protein has 141 amino acids with three cysteines located at positions 7, 86 and 95 (Figure 1), and thus is potentially capable of forming a number of intramolecular and intermolecular disulfide bonds. Pure r-hGMF-beta gave a single protein band of apparent molecular mass of 17 kDa on SDS-polyacrylamide gel electrophoresis under reducing and non-reducing conditions, confirming absence of any intermolecular disulfide linkages. The absence of intramolecular disulfide bond in r-hGMF-beta was confirmed by two methods. First, based on the observations of Clogston et al. (2) that the reduced and oxidized forms of r-HuIL-6 retained differently on reverse-phase HPLC, r-hGMF-beta was analyzed on a C18 reverse-phase column before and after treatment with a large excess of dithiothreitol (DTT). Results indicated identical retention times in each case (Fig. 2 A & B). Secondly, estimation of free thiol groups in native and denatured r-hGMF-beta by using Ellman's reagent (DTNB) (16) revealed presence of three SH-groups per mole protein (Table I). Thus, these results confirmed that r-hGMF-beta is in fully reduced form. The reduced form of r-hGMF-beta did not show any biological activity when tested on C6 glioma cells as described earlier (11). The biologically active form of r-hGMF-beta could be obtained only after oxidative refolding in the presence of a "redox" system consisting of reduced and oxidized glutathione in the presence of guanidine hydrochloride. The resultant oxidized form of the protein, presumably containing intramolecular disulfide bonded isoforms, has biological activity comparable to that reported earlier for natural bovine brain GMF-beta (11). Since GMF-beta has three cysteines, located at positions 7, 86 and 95, and can only form three intramolecular disulfide bonded isoforms as shown in the scheme (Figure 3), the biologically active "redoxed" r-hGMF-beta should contain these isomers plus the reduced form.

For the identification of biologically active form of GMF-beta, separation of these isoforms, which only differs in their molecular shape, was necessary. This was achieved by reverse-phase HPLC. For this purpose oxidative refolding was carried out, as described earlier (17), by treating a solution of r-hGMF-beta at a concentration of 250 μg per ml in 0.1M sodium phosphate containing 3M guanidine hydrochloride, pH 8, with 10mM reduced glutathione and 1mM oxidized glutathione at room temperature. The final concentration of guanidine hydrochloride was kept at 3 M which was found to be sufficient for complete denaturation of GMF-beta as determined by circular dichroic spectral analysis (results not shown). At various incubation

```
          7      10                              20                      30
S  E  S  L  V  V  C̲  D  V  A    E  D  L  V  E  K  L  R  K  F    R  F  R  K  E  T  N  N  A  A

                        40                           50                            60
I  I  M  K  I  D  K  D  K  R    L  V  V  L  D  E  E  L  E  G    I  S  P  D  E  L  K  D  E  L

                        70                           80              86            90
P  E  R  Q  P  R  F  I  V  Y    S  Y  K  Y  Q  H  D  D  G  R    V  S  Y  P  L  C̲  F  I  F  S

            95        100                           110                          120
S  P  V  G  C̲  K  P  E  Q  Q    M  M  Y  A  G  S  K  N  K  L    V  Q  T  A  E  L  T  K  V  F

                  130                      140
E  I  R  N  T  E  D  L  T  E    E  W  L  R  E  K  L  G  F  F    H
```

Fig. 1. Amino acid sequence of GMF-beta. The sequence was established for bovine brain GMF-beta by automated Edman degradation (microsequencing) and tandem mass spectrometry. Identical sequence was obtained for recombinant hGMF-beta, which was deduced from nucleotide sequence of the cDNA and verified by microsequencing of the first ten NH_2-terminal residues and by carboxylpeptidase degradation of the first four COOH-terminal residues. The one-letter abbreviations for the amino acids are: A, Ala; C, Cys; D, Asp; E, Glu; F, Phe; G, Gly; H, His; I, Ile; K, Lys; L, Leu; M, Met; N, Asn; P, Pro; Q, Gln; R, Arg; S, Ser; T, Thr; V, Val; W, Trp; and Y, Tyr. The three cysteine residues (at positions 7, 86 and 95) are underlined. (Adapted from ref. 13)

Table I. Sulfhydryl content of r-hGMF-beta

Treatment	SH-groups* (mole/mole)
None	2.7
DTT	2.8
Gu.HCl + DTT	2.8

*Determined with DTNB

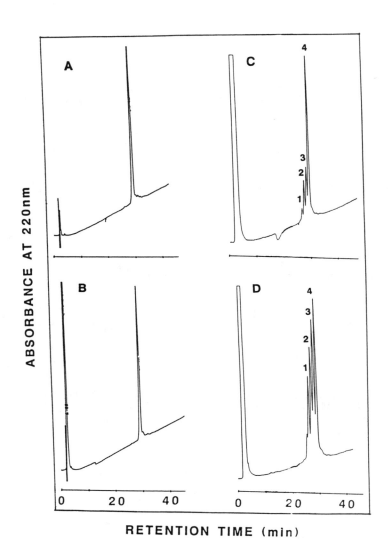

Fig. 2. Reverse-phase HPLC profile of r-hGMF-beta. (A) Pure r-hGMF-beta without treatment. (B) After treatment with 10 mM DTT in presence of 6M guanidine hydrochloride. Note identical retention time in each case. (C & D) Time course of oxidative refolding of r-hGMF-beta and separation of isoforms by reverse-phase HPLC. Pure r-hGMF-beta was dissolved in 0.1M sodium phosphate and incubated at room temperature with reduced and oxidized glutathione in the presence of guanidine hydrochloride, as described in the text, for 4 h (C) and 8 h (D). (Adapted from ref. 17)

times (shown in Figure 2) an aliquot was withdrawn and directly injected onto a uBondapack C18 column (3.9 mm X 30 cm, Waters). The solvents used were 0.1% trifluroacetic acid (solvent A) and 100% acetonitrile containing 0.1% TFA (solvent B). Samples were injected under initial column conditions of 30% solvent B at a flow rate of 1.5 ml/min. After 2 min, a linear gradient of 30-55% solvent B was established at a rate of 0.33% per min. Detection was made at 220 nm. A typical time course profile of refolding of r-hGMF-beta is shown in Figure 2 (C & D). It is evident from the results that r-hGMF-beta, under these conditions, was converted into four peaks. We speculated that these peaks corresponded to the reduced form plus three intramolecular disulfide bonded isoforms predicted from the number of cysteine residues. The "redoxed" GMF-beta was repeatedly injected onto the column and all the peaks were collected. Individual peaks were purified by rechromatography on HPLC under identical conditions and then analyzed by SDS-polyacrylamide electrophoresis under reducing and non-reducing conditions. The electrophoretic patterns indicated the presence of intact (17 kDa) GMF-beta in each peak and that all are monomers without any intermolecular disulfide linkage. Sulfhydryl determination with DTNB under denaturing conditions demonstrated that each peak has only one SH-group per mole , except peak #4 which contained three SH-groups per mole. Reduction with DTT followed by removal of the excess reagent and then estimation of the resulting thiol shows the presence of three SH-groups per mole in each peptide.

Based on the concept that the retention time of a polypeptide chain in reverse-phase HPLC increases with the degree of unfolding of the chain (2,4,18,19) we infered that HPLC peaks 1, 2, 3, and 4 corresponded to the isoforms so numbered, respectively, in Fig. 3. In order to determine which of the isoforms is biologically active, the eluted peaks from HPLC were assayed on C6 glioma cells for suppression of proliferation. Table II shows that only

Table II. Growth-inhibitory activity of isoforms of r-hGMF-beta

Sample	Cell Number (X 10^5)	Growth Inhibition (%)
Control	78.9 ± 2.8	0.0
Peak #1	74.9 ± 4.3	5.0
Peak #2	76.9 ± 4.2	2.5
Peak #3	20.1 ± 3.4	74.5
Peak #4	75.4 ± 5.2	4.4

Rat glioma line C6 (American Type Culture Collection) was seeded at 2 x 10^5 cells per well (in 24-well trays) in F12/DMEM (1:1) medium containing 5% (vol/vol) fetal calf serum. The cells were allowed to attach for 4 h and then exposed for 60 h to the GMF isoforms collected from the four HPLC peaks (Fig. 3). GMF protein was used at 100 ng/ml. Cell number was determined with a Coulter counter after trypsin treatment. Numbers 1-4 correspond to the HPLC peak numbers. Control cells were exposed to r-hGMF-β with redox treatment. Note that only peak 3 of the refolded isoforms was biologically active.

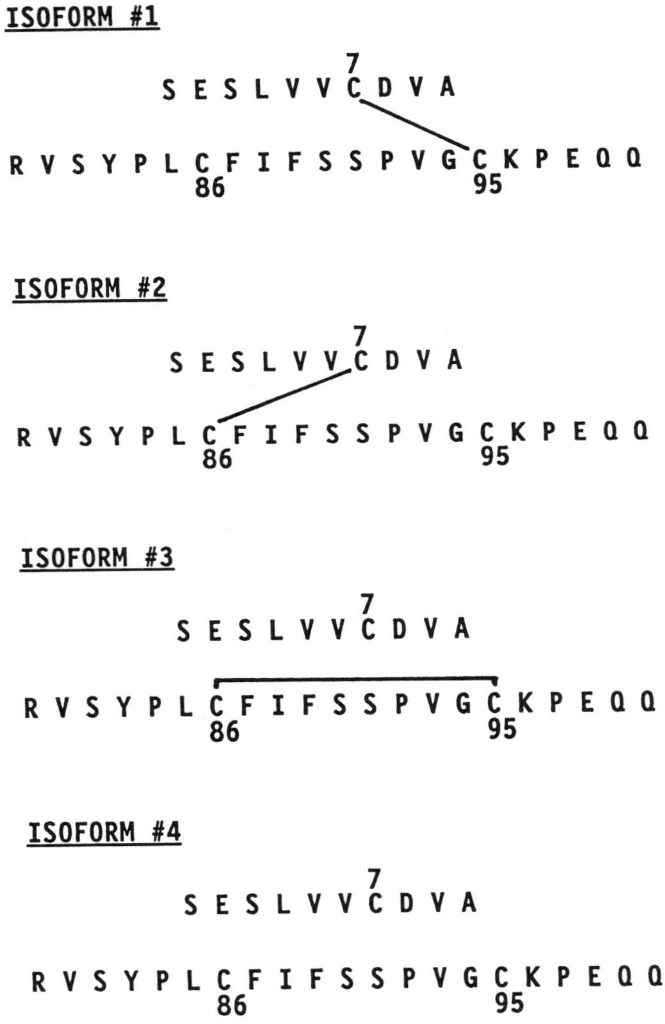

ISOFORM #1

```
                       7
          S E S L V V C D V A

R V S Y P L C F I F S S P V G C K P E Q Q
          86                    95
```

ISOFORM #2

```
                       7
          S E S L V V C D V A

R V S Y P L C F I F S S P V G C K P E Q Q
          86                    95
```

ISOFORM #3

```
                       7
          S E S L V V C D V A

R V S Y P L C F I F S S P V G C K P E Q Q
          86                    95
```

ISOFORM #4

```
                       7
          S E S L V V C D V A

R V S Y P L C F I F S S P V G C K P E Q Q
          86                    95
```

Fig. 3.   Predicted disulfide isoforms of r-hGMF-beta; showing three alternative intramolecular disulfide linkages.

peak #3 was biological activity. From this, we reasoned that the isoform with an intrachain disulfide bond between positions 86 and 95 is the biologically active species.

In this article, we demonstrate the feasibility of utilizing reverse-phase HPLC for the separation of disulfide-bonded r-hGMF-beta isoforms. The methodology permit us to predict the position of the disulfide linkage in each of the isoforms, and thus contributes to the understanding of structure-function relationship. However, we admit that the conclusion so obtained is only tentative. The inferred position of disulfide bond must be verified by collecting sufficient amounts of the isoforms for analysis by protein chemistry. But since HPLC columns can only handle a small amount of sample without sacrificing resolution, this approach appears laborious. A simpler alternative consists of replacing the cysteine residues, one at a time, with another amino acid, utilizing site-directed mutagenesis. Work along this direction is in progress.

Acknowledgments

We thank W.E. Franklin, B.D. Fink, J.L. Delp and C.A. Russo for technical assistance. This work was supported by the following grants to R.L.: Veterans Affairs Merit Review Award, National Science Foundation Grant BNS-8917665, and the Diabetes-Endocrinology Center Grant DK-25295.

Literature Cited

1.   Tsuji, T.; Nakagawa, R.; Sugimoto, N.; Fukuhara, K- I. *Biochemistry* 1987, *26*, 3129-3134.
2.   Clogston, C. L.; Boone, T. C.; Crandall, C.; Mediaz, E. A.; Lu, H. S. *Arch. Biochem. Biophys.* 1989, *272*, 144-151.
3.   Smith, M. C.; Cook, J. A.; Furman, T. C.; Occolowitz, J. L. *J. Biol. Chem.* 1989, *264*, 9314-9321.
4.   Morehead, H.; Johnston, P. D.; Wetzel, R. *Biochemistry* 1984, *23*, 2500-2507.
5.   Taniguchi, T.; Ohno, S.; Fujii-Kuriyama, Y.; Muramatsu, M. *Gene* 1980, *10*, 11-15.
6.   Lim, R.; Li, W. K. P.; Mitsunobu, K. Abstr. 2nd Ann. Meet. Soc. Neurosci., Houston, 1972, p. 181.
7.   Lim, R.; Mitsunobu, K.; Li, W. K. P. *Exp. Cell Res.* 1973, *79*, 243-246.
8.   Lim, R.; Mitsunobu, K. *Science* 1974, *185*, 63-66.
9.   Lim, R.; Hicklin, D. J.; Ryken, T. C.; Miller, J. F. *Dev. Brain Res.* 1987, *33*, 49-57.
10. Lim, R.; Hicklin, D. J.; Miller, J. F.; Williams, T. H.; Crabtree, J. B. *Dev. Brain Res.* 1987, *33*, 93-100.
11. Lim, R.; Miller, J. F.; Zaheer, A. *Proc. Natl. Acad. Sci. USA.* 1989, *86*, 3901-3905.
12. Lim, R.; Liu, Y.; Zaheer, A. *Dev. Biol.* 1990, *137*, 444-450.
13. Lim, R.; Zaheer, A.; Lane, W. S. *Proc. Natl. Acad. Sci. USA.* 1990, *87*, 5233-5237.
14. Kaplan, R.; Zaheer, A.; Jaye, M.; Lim, R. *J. Neurochem.* (In press).
15. Lim, R.; Zaheer, A. In *Methods Neurosci.*; Conn, P.M., Ed.; Vol. 6 (In press).
16. Ellman, G. L. *Arch. Biochem. Biophys.* 1959, *82*, 70-77.
17. Zaheer, A.; Lim, R. *Biochem. Biophys. Res. Comm.* 1990, *171*, 746-751.
18. Kunitani, M.; Johnson, D.; Snyder, L. R. *J. Chromatography* 1986, *371*, 313-333.
19. Lin, S. W.; Karger, B. L. *J. Chromatography* 1990, *499*, 89-102.

RECEIVED February 19, 1991

Chapter 7

Limited Proteolysis of Solvent-Induced Folding Changes of β-Lactoglobulin

M. Dalgalarrondo, C. Bertrand-Harb, J.-M. Chobert, E. Dufour, and T. Haertlé

Institut National de la Recherche Agronomique, LEIMA, BP 527, 44026 Nantes CDX 03, France

Strong hydrophobic core stabilizes the tridimensional structure of β-lactoglobulin molecule making its proteolysis with pepsin impossible since in aqueous solutions all cleavage sites are burried inside β-barrel. However, β-lactoglobulin molecule is subject to radical (but reversible) structural changes during the decrease of dielectric constant brought about by the addition of protic solvents such as alcohols. The analysis of far ultraviolet circular dichroic spectra of β-lactoglobulin indicates that its structure contains 52% of β-sheet in aqueous solutions. 65% of α-helix is observed in all studied β-lactoglobulin molecular species after addition of alcohol up to 50% v/v final concentration. These solvent induced structural changes can be followed by limited proteolysis. Cleavage of β-lactoglobulin with pepsin is triggered by the induced structural transformations and its speed and outcome is influenced by their extent. Differences in populations of produced peptides indicate the changes of folding intermediates present in the studied β-lactoglobulin solutions.

Recent developments in studies carried out on small proteins interacting with hydrophobic ligands have shed new light on the molecule of β-lactoglobulin [1 - 3]. Unexpected connectivities, analogies and new homologies have made this puzzling small protein even more interesting. Past studies of β-lactoglobulin have yielded a significant amount of information about the proteins in general. β-lactoglobulin was one of the first proteins isolated in its pure state and one of the first in which amino acid composition and sequence were established. Despite intensive studies there are still many intriguing questions which need to be answered about this protein and its potential as a target or as a tool of scientific research has not yet been exhausted.

It has been recently claimed that β-lactoglobulin belongs to the 'superfamily' of proteins involved in strong interactions with small volatile hydrophobic ligands [4 - 6]. Most of these proteins are also responsible for the inter and intracellular transport of the hydrophobic ligands, otherwise insoluble in a polar environment. Retinol binding protein [3], bilin binding protein [1], insecticyanin [2] and β-lactoglobulin [5, 7] are the best known proteins of this kind. Their tridimensional structures overlap in more than 95% and constitute a hydrophobic pocket inside of an eight-stranded β-barrel bordered on one side by an α-helix. It has been suggested that this kind of β-barrel structure might be a general structural device found in animal organisms used to trap and transport the small hydrophobic ligands.

0097–6156/91/0470–0086$06.00/0

In spite of relatively long acquired knowledge that β-lactoglobulin interacts strongly with retinol [8] the exact physiological role of β-lactoglobulin and its complex with retinol is unknown. This small protein is an abundant component of the milk (whey) of several mammals [6]. There is also evidence of intestinal translocation of this resistent to the acid proteolysis protein in humans. This can be deduced from reports about the presence of diet dependent β-lactoglobulin antigens in maternal milk [9]. All these observations indicate that β-lactoglobulin might pass across different membranery interfaces. Hence, it is important to investigate the behavior of this protein when it is subject to polarity changes as this may simulate, to some extent, the conditions met during its translocation through biological membranes. The structural changes of β-lactoglobulin under the influence of weak aprotic solvents in acidic pH have been studied [10, 11] previously. Analysis of ORD (optical rotatory dispersion) and IR (infrared radiation) data have indicated an important structural transformation of this protein. On the one hand, present methods allow further investigation of these phenomena. On the other hand, the dairy industry produces large amounts of β-lactoglobulin which could be engineered in order to bind and protect a wide range of lipophilic molecules such as flavors or drugs. Modification of β-lactoglobulin by enzymatic or chemicals treatments, as well as the changes of the dielectric constants of the medium, may be one of the possible ways for altering and broadening the binding specificity of this protein.

In contrast to spectroscopic measurements depicting only average structural changes, study of β-lactoglobulin limited proteolysis may give more topologically detailed informations about its folding changes [12]. The results of circular dichroism and limited proteolysis experiments aiming at the elucidation of alcohol-induced folding changes at acidic pH of β-lactoglobulin are presented and discussed in this paper.

Materials and Methods.

The preparation of β-lactoglobulin.   All chemicals used were reagent grade. β-lactoglobulin (BLG) variant B was obtained from homozygote cow's milk following the method of Mailliart & Ribadeau Dumas [13] and, as estimated from the high performance liquid chromatograms on a C_{18} column and polyacrylamide gel electrophoresis, it was found to be more than 95% pure.

Circular dichroism spectroscopy.   CD spectra were measured using a Jobin Yvon Mark III dichrograph linked to an Olivetti personal computer for data recording and analysis. Spectra were averages of 10 accumulated scans with substraction of the base line. The cylindrical cells used had a pathlength of 0.02 cm in the case of the far ultraviolet spectra (188-260 nm) and 1 cm in the 250-320 nm spectral region. All the spectra were taken at 20°C using β-lactoglobulin concentrations in the range of 20 − 30 μM. β-lactoglobulin concentration was determined spectrophotometrically using the molecular absorption coefficient ε_{278}=17600 in the calculations. β-lactoglobulin was disolved in 1 mM HCl or 0.1M sodium acetate buffer pH 3.0. The results are expressed in terms of molar ellipticity [Θ] (deg·cm²·dmole⁻¹).
The methods of Brahms & Brahms [14], Chen & Yang [15] or Chang et al. [16] were assayed in order to simulate the experimental spectra. Subsequently we used the Brahms' method which gave the best fit with the available β-lactoglobulin X-ray structural data, as described by Papiz et al. [5] and Monaco et al. [7].

Limited proteolysis with pepsin.   Proteolysis of β-lactoglobulin was performed in 20 mM sodium citrate buffer pH 2.5 with an addition up to 0, 20, 25, 30, 35, 40% (v/v) of ethanol. 300 μg of β-lactoglobulin in 200 μl (final volume) were hydrolyzed with 6 μg of pepsin (Sigma) - E/S ratio = 0.02. The proteolysis was performed in stabilized temperature of 20°C. Hydrolysis of each 200 μl aliquot was terminated by the addition of 300 μl of 0.2 M Tris-HCl buffer pH 8.0 at a given reaction time. The produced peptides were analyzed by HPLC on a 25x0.46 cm (-i.d.) Nucleosil C_{18} (porosity - 10 μm) column eluted with a linear gradient of 60% acetonitrile, 0.9‰ trifluroacetic acid in 1.1‰ trifluroacetic acid (starting solution), flow rate 1 ml/min, temperature 30°C, detection wavelength 214 nm. Identity of chosen peptides was deduced from their amino acid composition double-checked by the sequence analysis of first three N-terminal residues.

Results.

The influence of ethanol on the structural changes of β-lactoglobulin.
Far UV CD spectra of β-lactoglobulin are presented in Figure 1A. The addition of ethanol up to 50% v/v final concentration induces radical, but reversible, changes in circular dichroic spectra of β-lactoglobulin. In absence of ethanol the dichroic spectra exhibit sharp negative maxima in the aromatic region (Figure 1B) at 291 nm with strong shoulders at 284 nm. This pattern is characteristic of the tryptophan residues. In 50% ethanol (v/v), the two minima - at 284 and 291 nm disappear from the β-lactoglobulin spectrum.

The comparison of the fluorescence spectra of β-lactoglobulin in an aqueous solution and in 50% ethanol (v/v) (not shown) demonstrates that the maximum of the tryptophan fluorescence emission is shifted from 332 nm to 338 nm, respectively. Additionally, a concomitant increase in the maximum fluorescence intensity may be observed. Red shift of the emission maximum implies that under the influence of alcohol the tryptophan residues, which in aqueous solutions are sheltered in the hydrophobic interior of a protein molecule [17], become more exposed to a polar environment.

The analysis of far UV CD spectrum of β-lactoglobulin dissolved in aqueous solution, according to Brahms [14], demonstrates that it is characteristic of the proteins containing a large proportion of β-sheets, amounting to 52% for β-lactoglobulin (see Table I). This result is consistent with the conclusions of crystallographic analysis by Papiz et al. [5] and Monaco et al. [7]. Table I shows an estimate of the α-helix and β-sheet content using the Brahms method [14], based on the analysis of far UV CD spectra changes for β-lactoglobulin due to the increase in the ethanol concentration. In acidic 50% v/v ethanol, the observed UV CD spectrum of β-lactoglobulin is characteristic of proteins containing significant amounts of α-helix. Its analysis by Brahms method demonstrates the presence of 56% of α-helix, 10% of β-sheet and 34% of aperiodic segments (Table I).

β-strand <--> α-helix transition midpoints of BLG as a function of methanol, ethanol and 2-propanol concentrations.
The effect of increasing alcohol concentration on the molar ellipticity of the β-lactoglobulin solution measured at 191 nm is presented in Figure 2. It is an indirect tracer of the occuring β-strand <--> α-helix transition which culminates at around 50 % alcohol (v/v) concentration. It is immediately apparent that the shift of "β-strand half melting points" depends on the polarity of the used alcohol. The midpoints of β-lactoglobulin β-barrel "structure melting" are observed in 47% methanol (11.6M), 37% ethanol (6.35M) and 27% isopropanol (3.5M). Considering the bulk dielectric constant − (ε) values for alcohol/water solutions as given by Åkerlof [18] and Douzou [19], the midpoints of the observed structural transformation occur around dielectric constant $\varepsilon \approx 60$ for methanol, ethanol and 2-propanol. The molecule of β-lactoglobulin attains its maximal α-helix content when the (ε) value drops to about 50 in each alcohol studied.

Limited proteolysis of β-lactoglobulin.
It is known [20] that β-lactoglobulin is not digested in stomach and is almost integrally recovered at the entry of intestines.

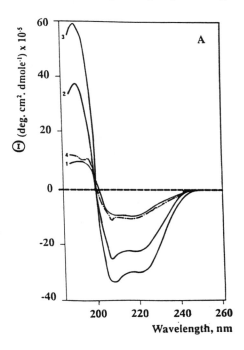

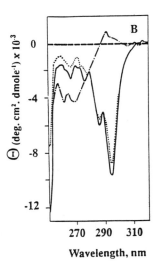

Figure 1 : Circular dichroism spectra of β-lactoglobulin in various aqueous (10^{-3}N HCl)/ethanol mixtures. A - Circular dichroism spectra in the far UV : (1) BLG in 10^{-3}N HCl, (2) BLG in 40% ethanol, (3) BLG in 50% acidic ethanol, (4) reversibility of the structural changes (BLG dissolved in 50% acidic ethanol is freeze dried and redissolved in 10^{-3}N HCl). B - Circular dichroism spectra in the aromatic UV region : BLG in 10^{-3}N HCL (——), in 20% (•••••) and in 50% acidic ethanol (– · —). BLG concentration was 18.2 μM.

Table I : Estimation of α-helix and β-sheet content in β-lactoglobulin using Brahms method as a function of ethanol concentration

% ethanol (v/v)	0	10	20	30	35	37.5	40	50
% H	7	10	10	13	24	39	46	56
% E	52	49	48	47	38	21	15	10
% R	41	41	42	40	38	39	39	34

% H : α-helix, % E : β-sheet and % R : aperiodic segments.

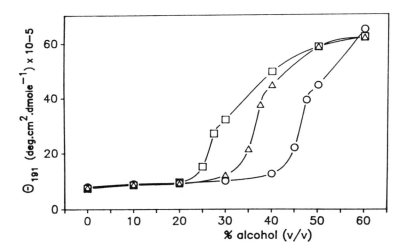

Figure 2 : Effects of methanol, ethanol and 2-propanol on β-lactoglobulin secondary structure. Changes in molar ellipticity at 191 nm of BLG dissolved in 10^{-3}N HCl as a function of methanol (o), ethanol (△) or 2-propanol (□) concentration. The results are expressed in terms of molar ellipticity [Θ] (deg·cm^2·dmole^{-1}).

Apparently, its hydrophobic core cages most of hydrophobic and aromatic amino acid side chains being otherwise sensitive to pepsic proteolysis.   Hence, it appeared plausible that alcohol induced structural changes might gradually expose the pepsin sensitive cleavage sites. The results of experiment supporting this hypothesis is presented in Figure 3. As it may be seen from HPLC chromatograms β-lactoglobulin which is not proteolyzed by pepsin *in vivo* can be neither hydrolysed *in vitro*, at pH 2.5 during 50 hrs of proteolysis time. Gradual addition of ethanol is inducing, however, an exponential boost of pepsic hydrolysis (represented in the Figure 3 as disappearence of the protein in time). This can be observed up to 35% of ethanol. The proteolysis in 40% ethanol is somewhat depressed by the prevailing inhibitory effect of alcohol. More detailed image can be perceived from the Figures 4A and 4B showing full HPLC chromatograms of the reaction products after 10 and 40 hours of pepsic hydrolysis of β-lactoglobulin in 25 and 40% of ethanol. At the first glance two groups of proteolytic products can be observed: - one of short more hydrophilic, preponderant at 25% alcohol and a second, of longer and more hydrophobic peptides, prevailing at 40% ethanol.

Discussion.

All the described experiments were carried out at pH 2 - 3 when β-lactoglobulin is monomeric and displays significant conformational stability. In alcohol, BLG CD spectrum shows the collapse of the β-sheet structure. The proteolysis data indicate that the changes of β-lactoglobulin secondary structure are preceded by earlier disorganization of its hydrophobic core. These structural transformations are paralleled by the change in the environment of at least one tryptophan, as might be deduced from analysis of the aromatic region of the CD and fluorescence spectrum. The changes of electrostatic charges, ionisation status, protein hydrophobicity and hydrogen bonds contribute to the breakdown of the β-barrel. The addition of alcohol decreases the dielectric constant of the aqueous solution and may: (i) induce the complete protonation of the aspartic and glutamic acid carboxylates and (ii) increase electrostatic repulsive interactions between the charges on the protein molecule in the solvent with a dielectric constant lower than water. Titrations by NaOH (0.2 N) of BLG (40 mg/ml) in aqueous solutions and in 50% ethanol indicate that acidic pK_a measured in BLG increases by about 1 unit from $pK_a = 3.7$ to $pK_a = 4.6$, respectively. This result agrees with one reported by Jukes & Schmidt [21] who showed that the pK_a of Asp and Glu side chains increase from 3.65 and 4.25 to 5.2 and 5.63 in 72% ethanol solution, respectively. The disappearance of the salt bridges and the overall change in the balance of the charges may contribute to a decrease in β-barrel stability.

It has been known for quite a long time that also several peptides can undergo conformational changes from an unordered or β-structure to an α-helix. For example, signal sequences of the unprocessed secretory proteins (transmembrane signal peptides) [22] are structurally unordered in aqueous solution. The interaction of these peptides with polar heads of membrane lipids induces them to adopt β-structure. After further penetration of the membrane and contact with its hydrophobic interior their structure becomes α-helical. In a similar process they become α-helical [22] upon interacting with non-polar solvents (trifluoroethanol, hexafluoroisopropyl alcohol). On the other hand, proteins are regarded as molecules with clearly defined ordered structures, which *a priori* cannot exhibit reversible conformational changes of great magnitude without imminent risk of denaturation. Structural studies by optical rotatory dispersion [10, 23, 24] have demonstrated, however, that some of the proteins - silk fibroin, ribonuclease and β-lactoglobulin, can undergo reversible conformational changes when dissolved in weakly protic solvents. Also, according to our CD observations the β-strand <--> α-helix transformation of β-lactoglobulin molecules under the influence of alcohols is reversible (Figure 1A). It should not be forgotten, however, that β-lactoglobulin is a secretory protein, and it is known as such

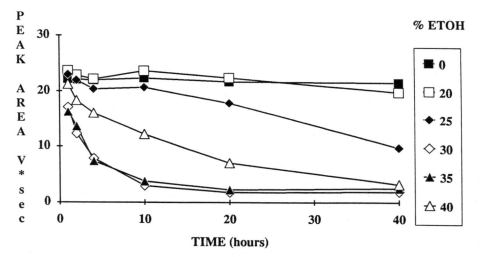

Figure 3 : Disappearence in time of β-lactoglobulin peak after proteolysis with pepsin as a function of ethanol concentration.

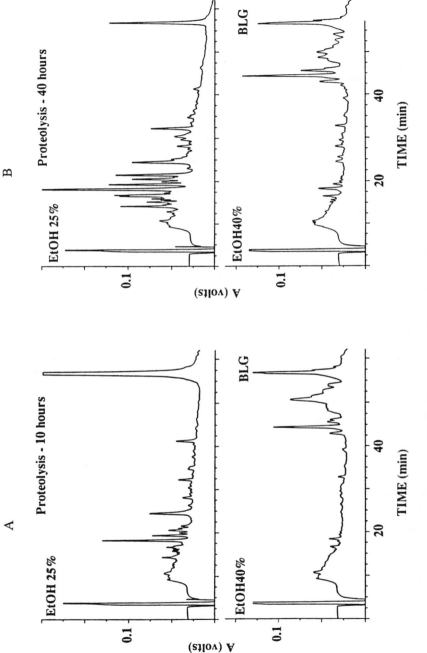

Figure 4 : HPLC of β-lactoglobulin pepsic peptides. A - 10 hrs , B - 40 hrs of reaction time. 25% above and 40% ethanol below.

to pass through the membranes of the endoplasmic reticulum. It has been shown that a conformational change is associated with the transport of β-lactoglobulin across the membrane [25]. The interaction of β-lactoglobulin with phospholipid bilayers - in phosphatidylcholine vesicles, increases the α-helix content of the protein, as determined by circular dichroism [26].

The physiological function of β-lactoglobulin, the major whey component of milk from cattle and other mammals is still far from being elucidated. The structural similarities between the retinol binding protein (RBP) and β-lactoglobulin, indicate that β-lactoglobulin could transport retinol or other hydrophobic molecules from the cow to its calf and evidence about the presence of the β-lactoglobulin receptor in the gut of very young calves [5] has been obtained. As β-lactoglobulin binds retinol tightly, one is tempted to postulate that the transformation of β-lactoglobulin (β-strands <--> α-helix), in an apolar medium, might explain the mechanism of retinol release and possibly also some of the transformations of this protein occuring during its translocation through cellular interfaces, such as membranes. This assumption is consistent with the results of Noy & Xu [27]. They report that, in a system containing retinol-RBP and lipid bilayers, the retinol spontaneously dissociates from retinol binding protein. So, the retinol carrier system appears to be in a dynamic equilibrium in which retinol can spontaneously move between binding proteins and lipid bilayers. Considering the strength of the known interactions, the shielding of all hydrophobic amino acid side chains by hydrophilic periphery of β-barrel (as witnessed by the failure of pepsin attack) and the hydrophobic nature of the ligands, any change in their binding simply seems unlikely, if not impossible, in the aqueous environment. Hence, it is very difficult to understand the mechanisms of the protein-ligand association/dissociation in a polar environment. In contrast, the observed β-strand<->α-helix transformation of β-lactoglobulin, may explain the possible mechanism of the delivery and binding of a variety of small hydrophobic ligands transported by other members of the same protein 'superfamily'. It might well explain the paradox arising from the strong binding by these proteins of the majority of hydrophobic ligands and the obvious need for their dissociation. The loading or unloading of the hydrophobic pocket may be intrinsic to the reversible processes of β-barrel collapse next to and inside of biological interfaces, when these proteins are fusing with or are secreted from the cellular membranes.

The analysis of limited proteolysis experiments presented in Figure 3 indicates that β-lactoglobulin structure is very compact and stable in aqueous solution at pH 2.5 since it is hiding efficiently all the hydrophobic side chains in the hydrophobic core of the β-barrel. The enhanced thermodynamic stability of β-lactoglobulin to thermal unfolding at low pH has been reported by Kella & Kinsella [28]. Also Novotny & Bruccoleri [29] and Parsell & Sauer [30] reported that structural stability is an important factor in proteolytic susceptibility dependent on enzyme accessibility and segmental mobility. As pepsic cleavage sites of β-lactoglobulin are well burried in its hydrophobic core they cannot be attacked by pepsin in aqueous solution. Consequently, the protein cannot be processed by this enzyme in regular physiological conditions. Initial addition of ethanol up to 20% doesn't exercise any significant influence. As it can be seen in Figure 3 further increase of ethanol content to 35% opens up most of pepsic cleavage sites and greatly accelerates the proteolysis. Quite surprisingly, as it shown in Figure 4A and 4B and partially confirmed by performed analysis, a number of relatively short hydrophilic peptides is produced at lower alcohol and few longer hydrophobic ones at higher ethanol concentrations. It looks like as at the beginning of studied polarity changes (20-30% of ethanol), when the spectral measurements still dont show any significant secondary structure changes (Figure 2), there might be tertiary structure tranformations and apparent from enhanced proteolysis - increased amino acid side chain rotation . In this respect, β-lactoglobulin conformation, observed around 20-30% ethanol might be similar to the "molten globule" state, suggested by Bychkova et al. [31] as a transitory conformation of globular proteins during their translocation through the membranes, conserving secondary while relaxing their tertiary structures.

It may be noticed (Figure 4) that the peptides produced in 40% ethanol are longer and more hydrophobic. At this value of dielectric constant, the plateau of the peptide backbone changes, as monitored by the CD measurements at 191 nm (see Figure 2), is reached and the maximum of α-helical structure is attained. Since the conformational freedom of amino acid side chains is greatly reduced in α-helices [32] many of the hydrophobic amino acid side chains may be hindered and consequently less exposed to the pepsin recognition on the surface of newly constituted α-helices. Hence, few observed long peptides are probably the products of pepsic cleavages at the inter-helical kinks or other non-helical rotationally less stable regions.

Acknowledgements. We would like to express our gratitude to Dr. L. Sawyer from the Department of Biochemistry, The University of Edinburgh Medical School for the tridimensional coordinates of β-lactoglobulin structure, to Dr. F. Toma from Service de Biochimie, CEA de Saclay for allowing us to use the circular dichrograph and to Dr. B. Ribadeau Dumas from INRA in Jouy-en-Josas for his stimulating discussions and critical comments.

The work presented in this paper has been funded by the Institut National de la Recherche Agronomique as part of the: "Study of the hydrophobic interactions of β-lactoglobulin" project.

Literature Cited.

1. Huber, R., Schneider, M., Epp, O., Mayr, I., Messerschmidt, A., Pflugrath, J. and Kayser, H. *J. Mol. Biol.* **1987**, 195, 423-434.
2. Holden, H.M., Rypniewski, W.R., Law, J.H. and Rayment I. *EMBO J.* **1987**, 6, 1565-1570.
3. Newcomer, M.E., Jones, T.A., Aqvist, J., Sundelin, J., Eriksson, U., Rask, L. and Peterson, P. *EMBO J.* **1984**, 3, 1451-1454.
4. Pervaiz, S., and Brew, K. *Science* **1985**, 228, 335-337.
5. Papiz, M.Z., Sawyer, L., Eliopoulos, E.E., North, A.C.T., Findlay, J.B.C., Sivaprasadarao, R., Jones, T.A., Newcomer, M.E. and Kraulis P.J. *Nature* **1986**, 324, 383-385.
6. Godovac-Zimmerman, J. *Trends in Biochem. Sci.* **1988**, 13, 64-66.
7. Monaco, H.L., Zanotti, G., Spadon, P., Bolognesi, M., Sawyer, L. and Eliopoulos, E.E. *J. Mol. Biol.* **1987**, 197, 695-706.
8. Futterman, S. and Heller, J. *J. Biol. Chem.* **1972**, 247, 5168-5172.
9. Monti, J.C., Mermoud, A.F. and Jolles,P. *Experientia* **1989**, 45, 178-180.
10. Tanford, C., De, P.K. and Taggart, V.G. *J. Am. Chem. Soc.* **1960**, 82, 6028-6035.
11. Townend, R., Kumosinski, T.F. and Timasheff, S.N. *J. Biol. Chem.* **1967**, 242, 4538-4545.
12. Jaenicke, R. and Rudolph, R. *in Protein Structure*, IRL Press, **1989**, Creighton, T. E. Ed., 191 -223.
13. Mailliart, P. and Ribadeau Dumas, B. *J. Food Sci.* **1988**, 53, 343-745.
14. Brahms, S. and Brahms, J. *J. Mol. Biol.* **1980**, 138, 149-178.
15. Chen, Y.H. and Yang, J.T. *Biochem. Biophys Res. Com.* **1971**, 44, 1285-1291.
16. Chang, C.T., Wu, C.S.C. and Yang, J.T. *Anal. Biochem.* **1978**, 91, 13-31.
17. Stryer, L. *Science* **1968**, 162, 526-533.
18. Åkerlöf, G. *J. Am. Chem. Soc.* **1932**, 54, 4125-4139.
19. Douzou, P. *in Cryobiochemistry* Acad. Press Inc. New York, **1977**, pp 27-45.
20. Yvon, M., Van Hille, I., Pélissier, J. P., Guilloteau, B., Toullec, R. *Reprod. Nutr. Develop.* **1984**, 24(6) 835-843.
21. Jukes, T.H. and Schmidt, C.L.A. *J. Biol. Chem.* **1934**, 105, 359-371.

22. Gierasch, L.M. *Biochemistry* **1989**, 28, 923-930.
23. Yang, J.T. and Doty, P. *J. Am. Chem. Soc.* **1957**, 79, 761-775.
24. Sage, H.J. and Singer, S.J. *Biochemistry* **1962**, 1, 305-317.
25. Mercier, J.C. and Gaye, P. *Ann. N.Y. Acad. Sci.* **1980**, 343, 232-251.
26. Brown, E.M., Carroll, R.J., Pfeffer, P.E. and Sampugna, J. *Lipids* **1983**, 18, 111-118.
27. Noy, N. and Xu, Z.-J. *Biochemistry* **1990**, 29, 3878-3883.
28. Kella, N.K.D. and Kinsella, J.E. *Biochem. J.* **1988**, 255, 113-118.
29. Novotny, J. and Bruccoleri, R.E. *FEBS lett.* **1987**, 211, 185-189.
30. Parsell, D.A. and Sauer, R. *J. Biol. Chem.* **1989**, 264, 7590-7595.
31. Bychkova, V.E., Pain, R.H. and Ptitsyn, O.B. *FEBS lett.* **1988**, 2, 231-234.
32. Piela, L., Nemethy, G. and Scheraga, H.A. *Biopolymers* **1987**, 26, 1273-1286.

RECEIVED February 6, 1991

Chapter 8

Folding and Aggregation of RTEM β-Lactamase

Pascal Valax and George Georgiou

Department of Chemical Engineering, University of Texas—Austin,
Austin, TX 78712

High levels of expression of the secreted protein RTEM β-lactamase in
Escherichia coli result in the formation of protein aggregates, or
inclusion bodies, in the periplasmic space. The formation of inclusion
bodies can be inhibited in cells grown in the presence of non-
metabolizable sugars such as sucrose. Earlier work has shown that
sucrose appears to exert a direct effect on the folding of β-lactamase
within the cell (*14*). In this study we have investigated the *in vitro*
renaturation and aggregation of β-lactamase from guanidine-HCl
solutions. The equilibrium folding transitions monitored by difference
spectroscopy, intrinsic fluorescence and activity measurements exhibited
some degree of non-coincidence indicating the existence of at least one
stable intermediate. Denaturation was not fully reversible at protein
concentrations in excess of 4 mg/ml and resulted in the formation of
visible aggregates. The extent of aggregation was greater when the
protein was first completely unfolded in buffers containing 2.0 M or
higher concentrations of guanidine-HCl, or in the presence of 5 mM
dithiothreitol. Addition of sugars caused a shift of the equilibrium
curves to higher guanidine-HCl concentrations and prevented the
aggregation of β-lactamase upon refolding. These results are consistent
with the aggregation inhibition observed *in vivo* in the presence of
saccharides.

The recent development of recombinant DNA technology has permitted the cloning of
foreign genes in microorganisms. In some cases the rate of production of the
recombinant protein exceeds 30% of the host total protein synthesis rate. Often, the
recombinant proteins accumulate in a misfolded conformation forming relatively large,
amorphous aggregates called inclusion bodies. Little is known about the mechanisms
of protein folding and aggregation in the cell. Recent studies have demonstrated that
interactions with intracellular components are critical for folding. For example,
chaperonins (or PCB, Polypeptide Chain Binding proteins) such as GroEL, GroES and
SecB (*1-7,10*) mediate proper folding of certain proteins, help maintain secretory

0097–6156/91/0470–0097$06.00/0

proteins in a conformation competent with membrane translocation and promote the correct assembly of oligomeric proteins. Folding catalysts such as protein disulfide isomerase (8-10) and prolyl cis-trans isomerase (10) are also believed to play an essential role. Despite the complexity of intracellular events, valuable insights on the folding pathway can often be drawn from *in vitro* experiments with purified proteins refolded from denaturant solutions. Based on such analogies as well as *in vivo* experimental evidence, Mitraki and King (11) proposed that the formation of inclusion bodies proceeds through a mechanism analogous to the aggregation of polypeptides during refolding from denaturant solutions. Thus, it was suggested that the aggregation of proteins *in vivo* most likely results from specific intermolecular interactions between exposed hydrophobic surfaces of a soluble kinetic intermediate in the folding pathway of the recombinant protein. In general, folding and aggregation can be considered as two competing reactions. The final yield of soluble, correctly folded protein is determined by the ratio of the rates of these two processes. Since the rate of polypeptide folding (for monomeric proteins) is first order with respect to protein concentration, whereas aggregation follows second or higher order kinetics, the formation of inclusion bodies should be favored at high intracellular protein concentrations arising from elevated expression levels.

Previous work in this laboratory has demonstrated that β-lactamase is a good model for the study of *in vivo* aggregation. This protein has been overexpressed in *E. coli* using an inducible *tac* promoter. Following translation, the precursor form of β-lactamase is translocated through the inner membrane and secreted into the periplasmic space. After the signal sequence has been cleaved, the released mature enzyme can either fold into the native conformation or aggregate in the periplasmic space forming inclusion bodies (12). Unlike the highly regulated cytoplasm, the periplasmic space is very sensitive to external conditions. Small molecular weight compounds can diffuse freely through the outer membrane. As a result, the fermentation conditions can directly influence the periplasmic space environment and therefore the folding and aggregation of secreted proteins. Addition of non-metabolizable sugars such as sucrose and raffinose has been shown to inhibit inclusion body formation (13). Bowden and Georgiou observed up to a ten-fold increase in the production of soluble native enzyme upon addition of sucrose in the fermentation medium. They also showed that the inhibition of inclusion body formation cannot be attributed to osmotic effects or changes in the protein synthesis rate (14). Instead, sucrose seems to act directly on the folding and aggregation of β-lactamase. In this study, experiments were designed to evaluate the effect of sucrose, as well as other factors such as pH, reduction of the disulfide bond and denaturant concentration on the extent of *in vitro* aggregation of purified β-lactamase.

Materials and Methods

Materials. Guanidine hydrochloride was purchased from International Biotechnologies Inc. (New Haven, CT). The purity of the lot was tested spectrophotometrically (15). Dithiothreitol (DTT) was purchased from Sigma.

General Methods. β-lactamase activities were determined spectrophotometrically using penicillin G as the substrate. The assay solution consisted of 0.5 g/l penicillin G in 50 mM potassium phosphate buffer, pH 6.5. All data are the average of two or three activity measurements. Enzyme concentrations were obtained from the absorbance at 281 nm using an extinction coefficient of $\varepsilon_{281}=29400$ $M^{-1}cm^{-1}$ (16,17). The molecular weight of the enzyme is 28899 g/mol. For some experiments, β-lactamase concentrations were also measured using the Bio-Rad binding dye assay with bovine serum albumin as the standard. Dialysis tubing (SPECTRUM, M.W. cutoff 10,000 daltons) was prepared by first boiling in 2% sodium bicarbonate, 1 mM EDTA for 10 min and then boiling for an additional 10 min in 1 mM EDTA (*18*). The tubing was extensively washed with distilled, deionized water before and after each boiling step and was stored at 4°C.

β-Lactamase Purification.   RTEM β-lactamase was produced by *E. coli* RB791:pTac11 grown at 37°C in M9 medium supplemented with 0.3 M sucrose. The culture was induced with 0.1 M isopropylthiolgalactoside (IPTG; 10^{-4} M final concentration) at an optical density O.D.$_{600}$ between 0.35 and 0.40. Under these conditions, most of the enzyme is released in the growth medium in stationary phase cultures (*13*). After 24 hours of cultivation, the cells were harvested by centrifugation (8,000xg for 8 min). Ammonium sulfate was first added to 30% of saturation and the precipitate was removed by centrifugation at 10,000xg for 40 min. Subsequently, the supernatant was saturated with ammonium sulfate and recentrifuged (10,000xg for 40 min.). The pellet was resuspended at room temperature in 300 ml of 20 mM Tris-HCl, pH 7.0 and dialyzed overnight at 4°C against 10 liters of the same buffer. It was then applied to a Waters QMA (quaternary methyl amine) preparative scale ion exchange HPLC column (particle size: 37-55 μm; pore size: 0.05 μm). The column was eluted with a linear gradient from 0 to 1.0 M NaCl in 20 mM Tris-HCl buffer, pH 7.8 developed over 45 min. The active fractions were pooled and the enzyme solution obtained was dialyzed against phosphate buffer at pH 7.0. 220 mg of β-lactamase were recovered from three 2-liter fermentations. β-lactamase was found to be more than 95% pure by SDS-polyacrylamide gel electrophoresis on 15% acrylamide gels. For long term storage, the enzyme was dissolved in 50 mM KH_2PO_4 buffer, pH 7.0, at a concentration of 1 mg/ml, was rapidly frozen in dry ice and kept at -70°C.

Folding Equilibrium Studies.   *E. coli* RTEM β-lactamase is a monomeric protein. Its amino acid sequence has been determined (*19*). It has one disulfide bond between the residues Cys[75] and Cys[121]. The presence of four tyrosines and four tryptophans allows the use of spectroscopic methods for the conformational characterization of the enzyme. In this study, the effect of denaturants on the unfolding of β-lactamase was determined from activity measurements, difference spectroscopy and fluorescence intensity measurements.

The effect of GuHCl on the enzymatic activity of β-lactamase was determined as follows: β-lactamase was dissolved in solutions containing different GuHCl concentrations. The protein concentration was 10^{-3} mg/ml for all experiments. All samples were incubated for three hours at room temperature to reach equilibrium. 500 μl of each sample was then mixed with 500 μl of phosphate buffer containing the same concentration of GuHCl and 1 mg/ml of penicillin G substrate. The activity was measured immediately as described above.

Difference spectra were recorded with a Varian DMS 200 double beam spectrophotometer by using the tandem cell technique (*20-22*). The protein concentration in all experiments was 0.5 mg/ml. Samples were equilibrated with the appropriate GuHCl concentrations for three hours at room temperature prior to absorbance measurement. The maximum difference in absorption in 6M GuHCl is observed at 286.5nm (Figure 1). All data presented were obtained by monitoring changes in the difference extinction coefficient at this wavelength.

A Perkin-Elmer LS5 fluorescence spectrophotometer was used for fluorescence intensity measurements. The excitation wavelength was set at 280 nm and the emission spectra were recorded between 310 and 480 nm. The protein concentration was 0.05 mg/ml in all experiments. The maximum difference in fluorescence intensity for the native and denatured enzyme was observed at 345 nm (Figure 2). Therefore, the

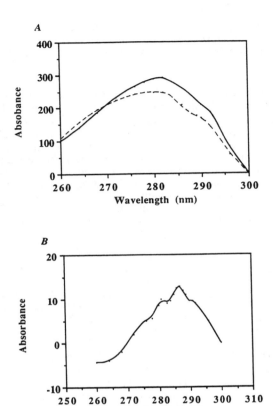

Figure 1. A. Absorption spectra of denatured (- -) and native (—) β-lactamase (0.5 mg/ml) in 50 mM phosphate buffer, pH 7.0 with and without 6 M GuHCl respectively. **B.** Difference spectrum of β-lactamase.

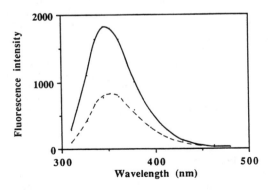

Figure 2. Fluorescence intensity spectra of native (—) and denatured (- - -) β-lactamase in 6 M GuHCl (0.05 mg/ml) in phosphate buffer, pH 7.0 . The excitation wavelength was 280 nm.

fluorescence intensity at this wavelength was used to monitor the extent of denaturation in the presence of intermediate concentrations of GuHCl.

Denaturation-Renaturation Procedure. Known amounts of β-lactamase were lyophilized and then dissolved in phosphate buffer, pH 7.0, containing various concentrations of GuHCl and dithiothreitol as described in the text. The samples were subsequently dialyzed against the same solution for three hours at room temperature in a PIERCE microdialyzer model 500 apparatus. The buffer was then changed and the protein allowed to refold for three more hours. In all experiments, the final GuHCl concentration was 0.02 M. The samples were subsequently centrifuged at 10,000 rpm for 20 min in an Eppendorf microcentrifuge at 4°C and the activity remaining in the supernatant was measured. The pellets were washed in 50 mM potassium phosphate, pH 7.0 and resuspended in the same buffer by vortexing. The suspension was then centrifuged as above and the enzyme activity of the supernatant was determined.

Results

Equilibrium Denaturation. A variety of different techniques can be employed to monitor protein conformational changes in the presence of denaturants. Activity measurements reflect the extent of alterations of the active site environment. However, enzyme activity measurements may be affected the presence of denaturant in the assay mixture. The denaturation curves obtained by this method are difficult to interpret and can only be taken as a first approximation of the unfolding transition. U.V. difference spectra indicate conformational changes by monitoring the degree of solvent exposure of aromatic amino-acid side chains. Finally, fluorescence intensity measurements can reveal the nature of the environment (polar, non-polar) of the four tryptophans of β-lactamase.

Folding equilibrium curves in the absence of sucrose are shown in Figure 3. The conformation of β-lactamase does not seem to undergo any perceptible change in the presence of up to 0.5 M GuHCl. On the other hand, no spectral change is observed above 1.25 M GuHCl, and therefore it is reasonable to assume that the transition is complete. All curves show a single step transition characteristic of a two-state process. However, Figure 3 reveals that the transitions detected by the three techniques are not completely coincident. This result indicates that at least one intermediate may be significantly populated at equilibrium. At 23°C, the unfolding of β-lactamase is completely reversible at concentrations at least as high as 2 mg/ml. Consequently, it is unlikely that the observed differences in the denaturation curves result from the protein concentrations used for the activity (1.0 μg/ml), fluorescence intensity (50 μg/ml), and difference spectroscopy (500 μg/ml) measurements. As pointed out by Pace (*23*), for multistate folding transitions, a two-state thermodynamic analysis can be used to calculate the lower limit of the Gibbs free energy of unfolding, ΔG_u. If y_N and y_D represent the values of the measured parameter for the native and denatured enzyme respectively, then for a given value y, of this parameter in the transition domain, the apparent equilibrium constant K_u can the be expressed as:

$$K_u = (y_N-y)/(y-y_D) \tag{1}$$

and the energy of unfolding ΔG_u is:

$$\Delta G_u = -RT\ln K_u \tag{2}$$

A simple linear regression was used to fit the data obtained by fluorescence measurements (Figure 4). $\Delta G_u(KH_2PO_4)$, the energy of unfolding of β-lactamase in

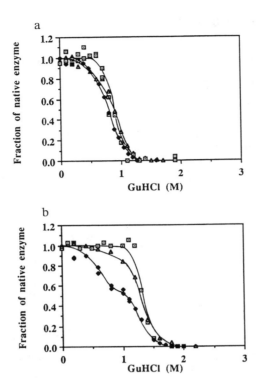

Figure 3. Denaturation equilibrium curves obtained from fluorescence intensity (▲), activity measurements (◆) and difference spectroscopy (▫) at 23°C, in the presence of different sucrose concentrations: a) no sucrose, b) 0.6 M sucrose (data for 0.3 M sucrose not shown). The solid lines are approximate fits for comparison of the experimental data.

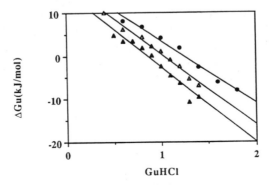

Figure 4. Thermodynamic treatment of folding equilibrium data obtained from fluorescence intensity measurements at 0 M (▲), 0.3 M (△) and 0.6 M (●) sucrose. Lines represent least square fits ($r \geq 0.98$ for all lines). The transition midpoints and values of the free energy change in the absence of denaturant are shown in Table I.

phosphate buffer, was obtained by linear extrapolation to zero concentration of denaturant.

Equilibrium denaturation experiments were also performed in buffers containing increasing concentrations of sucrose. The expected stabilization effect of sucrose against denaturation (*24-26*) was observed in all cases (Figure 4): the difference free energy values, estimated as described above, and midpoints of the unfolding transitions are listed in Table I. Increasing concentration of sucrose also lead to the appearance of an intermediate plateau in the transition monitored by activity measurements. It is not clear if this plateau represents the accumulation of an intermediate structure of β-lactamase with an altered active site or results from a solvent induced change in the enzyme kinetics.

Renaturation. The effects of i) reduction of the disulfide bond of β-lactamase, ii) renaturation buffer pH, iii) concentration of protein, iv) GuHCl in the denaturing solution and finally v) sucrose concentration on the reversibility of the unfolding transition were investigated.

To study the effect of the disulfide bond on the reversibility of the unfolding transition, β-lactamase was first unfolded in 2 M GuHCl in the absence or presence of 5 mM DTT. This concentration of the reducing agent represents a 5-fold molar excess with respect to the cysteine residues at the maximum protein concentration used (16 mg/ml). The unfolding buffer was thoroughly degased before the addition of DTT. Complete reduction of the disulfide bond of β-lactamase was confirmed by sulfhydryl titration using Ellman's reagent (data not shown). Subsequently, renaturation was performed by dialysis against potassium phosphate buffer, pH 6.0. As shown in Figure 5, protein aggregation is more pronounced when refolding is initiated with the reduced protein. For a 10 mg/ml protein solution, essentially all the activity is recovered without DTT, compared to approximately 80% with the reduced protein. In all cases, renaturation at high protein concentrations resulted in the formation of a visible precipitate. Virtually no activity could be recovered from the aggregates by dilution in phosphate buffer suggesting that precipitation is an irreversible process. Refolding of the reduced enzyme was also performed in the presence of 5 mM DTT in the renaturation buffer (Figure 5). The activity recovered in these experiments was equal to that obtained when the protein was renatured in the absence of DTT. The aggregates could be completely solubilized in high GuHCl concentrations (4 to 6 M) in absence of DTT and dilution or dialysis carried out at low protein concentrations resulted in full recovery of the enzymatic activity . SDS-polyacrylamide electrophoresis on 15% gels under non-reducing conditions showed that the refolded protein migrates as a single band. Furthermore, the refolded protein was subjected to gel filtration HPLC (Ultropack TSK G2000SWG, 21.5x300 mm, LKB Bromma) and eluted with potassium phosphate buffer, pH 7.0. Only a single, symmetric peak corresponding to the molecular weight of the native monomeric enzyme was observed (data not shown). These results clearly demonstrate that aggregation does not result from the intermolecular covalent cross-linking of cysteines.

The effect of pH on the reversibility of the unfolding transition was investigated by renaturing β-lactamase in potassium phosphate buffers at pH 6.0, 7.0 and 8.1 (Figure 6). As expected, refolding exhibited a strong dependence on pH. The maximum activity recovery was obtained at neutral pH. For a protein concentration of 12 mg/ml, 85% of active enzyme was recovered at pH 7.0, 75% at pH 6.0 and 60% at pH 8.1.

The purpose of this study is to draw analogies between *in vitro* observations and *in vivo* results on the aggregation of β-lactamase in *E. coli* (*12,13*). For this reason, it is important to conduct refolding experiments under conditions that approximate the *in vivo* environment as much as possible. For secreted proteins in *E. coli*, the formation of disulfide bonds occurs in the oxidative environment of the

Table I. Unfolding Parameters for β-Lactamase

Cosolvent	Transition Midpoint (M GuHCl)[¥]			ΔG (KH$_2$PO$_4$) (kJ/mol) [†]
	A	F	U	
1. No cosolvent	0.78	0.89	0.86	14.1±2.3
2. 0.3M Sucrose	0.87	1.06	....	16.8±1.8
3. 0.6M Sucrose	1.05	1.29	1.32	17.9±2.9

[¥]. Transition midpoints determined from A, activity measurement; F, fluorescence intensity and U, ultra violet absorption denaturation data.
[†]. Determined from fluorescence intensity data.

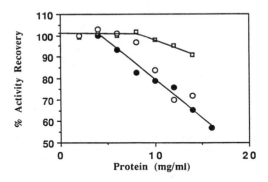

Figure 5. Percent of enzymatic activity recovered after renaturation by dialysis at 23°C. The protein was unfolded in 2.0 M GuHCl in 50 mM potassium phosphate buffer, pH 7.0 in the presence (●) and absence (◻) of 5 mM DTT. It was subsequently refolded in 50 mM potassium phosphate buffer, pH 6.0. The reduced enzyme was also dialyzed against 50 mM potassium phosphate, 5 mM DTT, pH 6.0 (○). The final GuHCl concentration was 0.02 M.

periplasmic space, concomitant with folding. Furthermore, the periplasm is thought to be at a pH lower than that of the medium due to the establishment of a Donnan equilibrium across the outer membrane (27). To account for these conditions, subsequent refolding experiments were conducted with a reduced protein at a pH of 6.0.

Native β-lactamase was incubated in GuHCl concentrations ranging from 0 to 6 M. The protein was then refolded, at a constant concentration of 10 mg/ml, in potassium phosphate buffer, pH 6.0. Irreversibility of the unfolding process was observed above 0.5 M GuHCl initial concentration (Figure 7). Between 0.5 and 2 M, the percentage of activity recovery decreased linearly from 100 to 80%. Unfolding in higher guanidine-HCl concentrations did not increase the recovery of active protein upon dialysis.

Sucrose has been shown to stabilize the native conformation of proteins and affect the folding kinetics (24-26, 28). Moreover, being a viscosity enhancer, sucrose decreases the rate of diffusion dependent processes such as aggregation. The addition of sucrose in the renaturation buffer should therefore lead to higher recovery of the active protein. The effects of 0.15 M and 0.3 M sucrose in the refolding buffer (potassium phosphate, pH 6.0) on protein aggregation were investigated (Figure 8). At both concentrations, the presence of sucrose in the refolding buffer inhibited aggregation. For a protein concentration of 16 mg/ml, about 60% of the activity was recovered without sucrose, 70% in 0.15 M sucrose and 85% in 0.3 M sucrose.

Discussion

The denaturation equilibrium experiments described in this paper show that β-lactamase becomes unfolded in the presence of more than 1.25 M guanidine-HCl. The non-coincidence of the difference spectra and fluorescence emission at guanidine-HCl concentrations between 0.3 and 0.75 M suggests the presence of an intermediate which is populated at equilibrium.

All the refolding equilibrium experiments showed that denaturation is reversible up to a certain protein concentration. Refolding by dialysis of relatively concentrated protein solutions leads to irreversible aggregation. It was demonstrated that intermolecular disulfide cross linking is not involved in the aggregation mechanism: β-lactamase contains a single disulfide bond linking Cys[75] and Cys[121] which is not required for full enzyme activity (29) or proper folding of the protein (30). On the other hand, there is some evidence that the formation of the disulfide bond is important for the stability of the enzyme at elevated temperatures (29). The results presented in this study suggest that the disulfide bond participates indirectly in the irreversible aggregation of β-lactamase. Hydrophobic interactions are usually the predominant driving force for protein association (31-35). Recent studies have demonstrated that *in vitro* aggregation results from the intermolecular interaction between specific exposed hydrophobic surfaces of partially folded species. Assuming that the *E. coli* β-lactamase and the homologous *S. aureus* penicillinase have similar structures (36), the disulfide bond would be expected to link two α-helices called α-2 (between residues 71 and 82) and α-4 (residues 119 to 125). The formation of the disulfide bond is likely to align the two helices and optimize the packing of hydrophobic domains. In the reduced protein, the two helices may be allowed to interact intermolecularly, thus enhancing aggregation.

London et al. (32) showed that the aggregation of the *E. coli* tryptophanase is initiated by the interaction between specific hydrophobic regions of a partially folded species. This species can be stabilized by intermediate GuHCl concentrations. Refolding from these GuHCl concentrations resulted in maximum aggregation and the appearance of a trough in the renaturation curves. Although no such minimum was

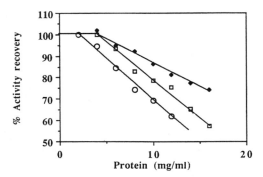

Figure 6. Effect of buffer pH on the activity recovery upon renaturation from 2 M GuHCl, 5 mM DTT at 23°C. (◆): pH 7.0; (□): pH 6.0; (○): pH 8.1. The final GuHCl concentration was 0.02 M.

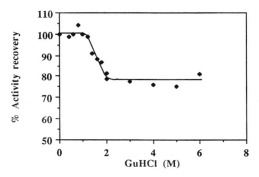

Figure 7. Percent of enzymatic activity recovered upon renaturation of 10 mg/ml of β-lactamase unfolded in different concentrations of GuHCl. The unfolding buffer contained 5 mM DTT in 50 mM potassium phosphate, pH 7.0 in addition to the denaturant. The protein was refolded by dialysis against 50mM potassium phosphate, pH 6.0 at 23°C. The final GuHCl concentration was 0.02 M for all experiments.

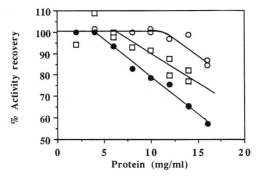

Figure 8. Effect of sucrose concentration on the renaturation of β-lactamase. Samples with different concentrations of protein were unfolded in 2.0 M GuHCl, 5 mM DTT and renatured by dialysis in potassium phosphate buffer, pH 6.0 (0.02 M final GuHCl concentration) with 0 M (●), 0.15 M (□) and (○) 0.3 M sucrose

observed with β-lactamase at 23°C (Figure 6), preliminary data indicate that a trough in the renaturation curve can be observed at higher temperatures. Mitraki et al. (*34*) have presented evidence suggesting that the entropy loss upon aggregation is most likely compensated by hydrophobic interactions which exhibit a strong temperature dependence. They showed that the aggregation of phosphoglycerate kinase (PGK) is inhibited at low temperatures at which hydrophobic interactions are not favored. The PGK renaturation curves showed a single step transition at 4°C and a minimum in protein recovery for intermediate concentrations of denaturant at 23°C.

Even though hydrophobic interactions have been shown to play a dominant role in the folding and aggregation of proteins, other interactions including electrostatic effects between between polar groups may also be important. Variations in pH alter the charge of polar amino-acid side chains. Because hydrophobic interactions are probably the main driving force for folding and aggregation, changes in the protein electrostatic properties may not significantly alter the conformations of critical folding intermediates but nevertheless may modify their solubilities making them prone to aggregation (*35*). We observed that the reversibility of β-lactamase unfolding is somewhat affected by the pH of the renaturation buffer. A similar, although more dramatic dependence of aggregation on pH has been observed with γ-interferon (*35*).

Sucrose and other sugars have been shown to stabilize proteins in their native conformations by inducing preferential hydration of the polypeptide chain (*24-26*). Sucrose has also been shown to increase the apparent refolding rate of the α-subunit of *E. coli* tryptophan synthase. The enhancement in the folding rate of this protein was first attributed to the stabilizing effect of sucrose. However, in a more recent study, Hurle et al.(*28*) developed a method that can take into account changes in equilibrium due to the addition of cosolvents. They found that the diffusion-dependent, intramolecular domain association during the refolding of tryptophan synthase was decelerated in presence of sucrose due to the increased viscosity. This result suggests that intermolecular domain association leading to aggregation may also be inhibited by sucrose. Our results demonstrated that the addition of low concentrations of sucrose to the renaturation buffer inhibits the formation of β-lactamase aggregates. Further studies to determine the precise mechanism through which sucrose affects β-lactamase folding and aggregation are under way.

In vitro and *in vivo* folding and aggregation are believed to follow identical pathways. The results obtained in this study can be used to interpret *in vivo* observations. The cytoplasm of *E. coli* is a relatively reducing environment that is believed to prevent the formation of disulfide bonds. The presence of the leader sequence retards the folding of β-lactamase in the cytoplasm, possibly with the help of chaperonins. The protein is therefore translocated across the inner membrane in a non-native, reduced form. The disulfide bond is subsequently formed in the periplasmic space, in which the concentration of dissolved oxygen is the same as in the growth medium. Sucrose is a non-metabolizable sugar which can diffuse through the outer membrane into the periplasmic space, but remains excluded from the cytoplasm. Addition of sucrose in the fermentation broth was shown to inhibit the formation of β-lactamase inclusion bodies (*13*). We have shown that the aggregation of β-lactamase during the refolding from GuHCl solutions is also prevented by the presence of sucrose. Thus, the effect of sucrose on *in vivo* aggregation is most likely related to changes in the folding transition rather than physiological alterations within the cell.

Acknowledgements.

We are grateful to L. Davidson and B. G. Kitto for their help with HPLC. We also thank Gregory Bowden and François Baneyx for reading the manuscript and Daniel

Thomas for his assistance in the lab. This work was supported by grant CBT 86-57971.

Literature Cited.

1.   Rothman, J. E. *Cell* **1989**, *59*, 591-601
2.   Ellis, R. J.; Hemmingsen, S. M. *TIBS* **1989**, *14(8)*, 339-342
3.   Lubben, T. H.; Donaldson, G. K.; Vitanen, P. V.; Gatenby, A. A. *The plant cell* **1989**, *1*, 1223-1230
4.   Goulobinoff, P.; Christeller, J. T.; Gatenby, A. A.; Lorimer, G. H. *Nature* **1989**, *342*, 884-889
5.   Kusukawa, N.; Yura, T.; Ueguchi, C.; Akiyama, Y.; Ito, K. *The EMBO Journal* **1989**, *8(11)*, 3517-3521
6.   Van Dyk, T. K.; Gatenby, A. A.; LaRossa, R. A. *Nature* **1989**, *342*, 451-453
7.   Phillips, G. J.; Silhavy, T. J. *Nature* **1990**, *344*, 882-884
8.   Lang, K.; Schmid, F. X.; Fischer, G. *Nature* **1987**, *329*, 268-270
9.   Fischer, G.; Wittmann-Liebold, B.; Lang, K.; Kiefhaber, T.; Schmid, F. X. *Nature* **1989**, *337*, 476-478
10.   Fischer, G.; Schmid, F. X. *Biochemistry* **1990**, *29(9)*, 2205-2212
11.   Mitraki, A.; King, J. *Bio/Technology* **1989**, *7*, 690-697
12.   Georgiou, G.; Telford, J. N.; Shuler, M. L.; Wilson, D. B. *Applied and environmental microbiology* **1986**, *52(5)*, 1157-1161
13.   Bowden, G. A.; Georgiou, G. *Biotechnology progress* **1988**, *4(2)*, 97-101
14.   Bowden, G. A.; Georgiou, G. *J. Biol. Chem.* **1990**, *265(28)*, 16760-16766
15.   Nozaki, Y. In *Methods in Enzymology*, Hirs, C. H. W.; Timasheff, S. N. Eds.; Academic Press, NY, 1972, vol. 26; 43-50
16.   Fischer, J.; Belasco, J. G.; Khosla, S.; Knowles, J. R. *Biochemistry* **1980**, *19*, 2895-2901
17.   Sigal, I. S.; DeGrado, W. F.; Thomas, B.J.; Petteway, Jr. S. R. *The Journal of Biological Chemistry* **1984**, *259(8)*, 5327-5332
18.   Maniatis, T.; Fritsch, E. F.; Sambrook, J. *Molecular Cloning, A Laboratory Manual*; Cold Spring Harbor Laboratory: Cold Spring Harbor, NY, 1982; 456
19.   Sutcliffe, J. G. *Proc. Natl. Acad. Sci. USA* **1978**, *75(8)*, 3737-3741
20.   Robson, B.; Pain, R. H. *Biochem. J.* **1976**, *155*, 325-344
21.   Schmid, F. X. In *Protein Structure, A Practical approach*; Creighton, T. E., Ed.; IRL Press, 1989, 251-285
22.   Donovan, J. W. In *Methods in Enzymology ;* Hirs, C. H. W.; Timasheff, S. N., Eds; Academic Press, NY, 1973, vol. 27; 497-525
23.   Pace, C. N. *CRC Critical Reviews In Biochemistry*; 1975; 1-38
24.   Lee, J. C.; Timasheff, S. N. *J. Biol. chem.* **1981**, *256(14)*, 7193-7201
25.   Arakawa, T.; Timasheff, S. N. *Biochemistry* **1982**, *21*, 6536-6544
26.   Arakawa, T.; Timasheff, S. N. In *Protein Structure, A Practical approach*; Creighton, T. E., Ed.; IRL Press, 1989, 331-345
27.   Stock, J. B.; Rauch, B.; Roseman, S. *J. Biol. Chem.* **1977**, *252(21)*, 7850-7861
28.   Hurle, M. R.; Michelotti, G. A.; Crisanti, M. M.; Matthews, C. R. *Proteins: Structure, Function and Genetics* **1987**, *2*, 54-63
29.   Schultz, S. C.; Dalbadie-McFarland, G.; Neitzel, J. J.; Richards, J. H. *Proteins: Structure, Function and Genetics* **1987**, *2*, 290-297
30.   Laminet, A. A.; Plückthun, A. *The EMBO Journal* **1989**, *8(5)*, 1469-1477
31.   Teipel, W. J.; Koshland, Jr. D. E. *Biochemistry* **1971**, *10(5)*, 792-805
32.   London, J.; Skrzynia, C.; Goldberg, M. E. *Eur. J. Biochem.* **1974**, *47*, 409-415

33.   Zettlmeissl, G.; Rudolph, R.; Jaenicke, R. *Biochemistry* **1979**, *18(25)*, 5567-5575
34.   Mitraki, A.; Betton, J. M.; Desmadril, M.; Yon, J. M. *Eur. J. Biochem.* **1987**, *163*, 29-34
35.   Mulkerrin, M. G.; Wetzel, R. *Biochemistry* **1989**, *28*, 6556-6561
36.   Herzberg, O.; Moult, J. *Science* **1987**, *236*, 694-701

RECEIVED February 6, 1991

Chapter 9

Role of Chaperonins in Protein Folding

Pierre Goloubinoff[1], Anthony A. Gatenby[2], and George H. Lorimer[2]

[1]Division of Biochemistry and Molecular Biology, University
of California—Berkeley, Berkeley, CA 94720
[2]Molecular Biology Division, Central Research and Development, E. I.
du Pont de Nemours and Company, Experimental Station,
Wilmington, DE 19880–0402

The chaperonin proteins, groEL and groES belong to a ubiquitous sub-family of heat-shock molecular chaperones, found in prokaryotes and in eukaryotic organelles. The chaperonins assist the folding of nascent, organelle-imported or stress-destabilized polypeptides. Using the photosynthetic enzyme Rubisco as a folding substrate, we demonstrate *in vitro* that this process requires firstly the formation of a binary-complex between an unstable Rubisco folding-intermediate with the oligomeric form of groEL. The second step, resulting in the formation of active Rubisco, consists of an ATP- and K^+-dependent discharge of the groEL-Rubisco binary complex. GroES is required as a coupling factor between the ATP hydrolysis and the Rubisco refolding reactions which take place on the groEL oligomer. At 25°C, chaperonins assist Rubisco refolding by preventing its aggregation and not by rescuing misfolded proteins.

The dictum that proteins fold into their final form in cells by a spontaneous event using information solely in the primary sequence is being challenged by recent discoveries. For numerous cellular processes, it is becoming apparent that there is a requirement for a class of auxiliary proteins whose proposed function is to govern the correct folding of other polypeptides. One name suggested for this family of proteins is molecular chaperones. A molecular chaperone is a protein characterized by its transient association with nascent, stress-destabilized or translocated proteins. This association prevents biologically unproductive phenomena and promotes the correct folding and assembly of target protein complexes into biologically active forms. The molecular chaperone is not a part of the final protein complex.

The term "molecular chaperone" is not novel and was first introduced (*1*) to describe the transient association of nucleoplasmin with the histones during the formation of nucleosomes. Subsequently, a similar transient association was found in the chloroplast of higher plants between nascent Large subunits (LSU) of the enzyme Ribulose bisphosphate carboxylase (Rubisco) and a 800kDa protein complex composed of 12-14 60kDa subunits (*2*). The nascent chloroplast-synthesized LSU was found to bind this complex in a non-covalent manner and in sub-stoichiometric amounts.

0097–6156/91/0470–0110$06.00/0

At first glance, the notion that the assistance of a molecular chaperone may be required for the folding and/or assembly of other proteins appears to be at variance with the work of Anfinsen (*3*), who demonstrated that bovine pancreatic ribonuclease can be denatured and consequently renatured *in vitro*, in the absence of other co-factors. This experiment has been repeated since with many other proteins, including Rubisco (*4*): thus, it has been assumed that the primary sequence is able and sufficient to direct the correct self-folding of all proteins into their functional tertiary structure.

There are, however, many proteins for which the conditions for *in vitro* self-refolding, have not yet been found or are not within physiological ranges. The dimeric form of Rubisco from the purple bacterium *Rhodospirilum rubrum* is a typical example of a protein whose spontaneous refolding, after a treatment in 8M urea, 6M guanidium-HCl or acid (pH 3.0), is temperature dependent (*4*). It is observed that, although 100% of active Rubisco may be recovered after a long incubation at 10°C, less that 1% can be recovered at 25°C. The remaining 99% are found in the form of an insoluble aggregate. Thus, there exists in the case of Rubisco and probably of many other proteins, an alternative pathway of aggregation, which competes at physiological temperature with the spontaneous, biologically productive folding pathway. At 25°C, the pathway leading to aggregates is believed to be caused by the improper interaction of solvent-exposed hydrophobic surfaces between unstable folding intermediates (*5*). This chapter focuses on the role of a sub-family of molecular chaperones named chaperonins (*6*), in the prevention of protein aggregation.

The Association Of Rubisco With The Chaperonins. During the last decade, evidence has accumulated that subunits of the photosynthetic enzyme Rubisco form a transient binary complex with another distinct protein: the Rubisco subunit binding protein (*7-9*). The significance of this transient complex remained obscure, until it was demonstrated that the Rubisco subunit binding protein and the groEL protein from *Escherichia coli*, shared a high degree of sequence homology (*10*). Genetic evidence (*11-14*) had previously implicated groEL in the assembly of bacteriophage lambda heads and T4 phage tails. Thus, a family of homologous proteins, termed "chaperonins", was associated with the formation of a diverse array of complex oligomeric structures.

The GroE Proteins Promote The Assembly Of Rubisco In *E.coli*.The finding of a high homology of sequences between the *E. coli* groEL and the Rubisco subunit binding protein suggested to us that groEL might also be involved in the folding and/or assembly of recombinant Rubisco in *E. coli*.

Gatenby *et al.* (*15*) had previously shown that a L8S8 Rubisco from the cyanobacterium *Anacystis nidulans* could be synthesized and correctly assembled into an active form in *E. coli*. To demonstrate the involvement of the chaperonin proteins in the process, we explored the influence of groEL and its co-transcribed groES chaperonin, on the folding and assembly process of the cyanobacterial L8S8 Rubisco (*16*). We constructed two sets of compatible plasmids, able to co-exist in the *E. coli* cell: One set of plasmids carried a pMB1 replicon minus or plus the genes for *A. nidulans* Rubisco *rbc*L+S (Figure 1, lanes 1 or 2-4 respectively). The second set of plasmids carried a p15A replicon plus (Figure 1, lanes 2, 3), or minus the *gro*ES+L genes (Figure 1 lane 1). In lane 4 of Figure 1, groES was inactivated by a frameshift mutation.

Rubisco activity was quantified in extracts of cells carrying various combinations of the two sets of plasmids. The amounts of soluble active Rubisco increased 4-5 fold when both the groEL and groES proteins were over-produced and co-expressed with the Rubisco large and small subunits. (Figure 1, lanes 3). When only groEL (and not groES) was over-produced, the solubility and activity of Rubisco remained low, at the levels limited by the chaperonins coded by the host chromosome (Figure 1, lane 4 and 2 respectively). However, an estimate by western blot analysis of the total amount of Rubisco large subunits present in the cell extracts, revealed no marked differences between the various treatments (data not show). Thus, the differences of Rubisco activity are not due to variations in the transcription, translation or degradation regimes of the Rubisco polypetides. The successful formation of an active L8S8 Rubisco enzyme in *E. coli*, rather depends upon a post-translational event (folding and/or assembly), determined by the levels of both groEL and groES chaperonin proteins in the cell.

This experiment emphasizes the key role played by the groES protein and suggests that a groES equivalent exists in chloroplast as well as in mitochondria. It also stresses the ubiquitous nature of the chaperonins and their lack of specificity. Obviously, groEL and groES do not exist in *E. coli* in order to assemble Rubisco, a protein not normally found in *E. coli*. Beside phage head and tail assembly (*11-14*), the groE operon is genetically associated with several other functions such as DNA and RNA replication (*17,18*) and amino acid synthesis (*19*). In yeast, conditional thermosensitive mutants in the function of the mitochondrial chaperonin, display large pleiotropic deficiencies due to unsuccessful targeting and assembly of many imported polypeptides (*20*). In addition, many proteins imported into the chloroplast are found associated with the chloroplast chaperonin (*21*). It appears that the rules that govern the interaction of the chaperonin with its multiple substrates must be of a very general nature.

The *In Vitro* Reconstitution Of Active Rubisco. To explore the molecular mechanism by which Rubisco assembly is promoted by the chaperonins, an *in vitro* system was developed. We chose the dimeric, simplified form of the *Rhodospirilum rubrum* Rubisco as our substrate, to be denatured with either 8M urea, 6M guanidium-HCl or acid treatment. The *E. coli* chaperonins, now referred to as cpn60 (groEL) and cpn10 (groES) were purified to homogeneity (*22*).

At $25^{\circ}C$, upon the removal (by dilution) of the denaturant, Rubisco forms an inactive aggregate that is unable to enter a non-denaturing gel (Figure 2, lane 2). However, if denatured Rubisco is diluted in a solution containing $[cpn60]_{14}$, a stable soluble binary complex is formed between $[cpn60]_{14}$ and Rubisco (Figure 2, lane 3). This complex is rather stable and can withstand gel electrophoresis. Note that $[cpn60]_{14}$ displays a specific affinity for denatured Rubisco, since no interaction is apparent between $[cpn60]_{14}$ and dimeric native Rubisco (Figure 2, lane 10). The addition of Mg-ATP or, alternatively, only $[cpn10]_7$ neither dissociates the $[cpn60]_{14}$^Rubisco complex, nor renders Rubisco enzymatically active (Figure 2, lanes 4 and 6, respectively). Only when both Mg-ATP and $[cpn10]_7$ are concomitantly added, is the $[cpn60]_{14}$^Rubisco complex being discharged and active folded dimeric Rubisco reformed (Figure 2, lane 7).

Chaperonins have to interact with unfolded Rubisco (Rubisco-U) in a specific order (table 1). The presence of $[cpn60]_{14}$ (not $[cpn10]_7$) is absolutely necessary as soon as the denaturant has been removed by dilution (Table 1, lane 1-4, versus 5). Pretreatment of $[cpn60]_{14}$ with Mg-ATP reduces the efficiency of the chaperonin to reconstitute active Rubisco (Table 1 lane 2 and 1) due to partial disruption of the $[cpn60]_{14}$ oligomer (Figure 2C lanes 1, 4, 7). Thus, the observed decrease in reconstitution ability may be assigned to the reduction of the active $[cpn60]_{14}$ form that binds and stabilizes the Rubisco folding intermediate (Rubisco-I) (see also Figure

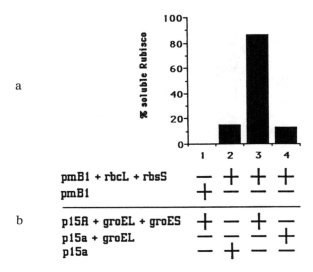

a

b

	1	2	3	4
pmB1 + rbcL + rbsS	−	+	+	+
pmB1	+	−	−	−
p15A + groEL + groES	+	−	+	−
p15a + groEL	−	−	−	+
p15a	−	+	−	−

Figure 1: *In vivo* assembly of Rubisco in *E. coli*. The A. nidulans rbcL and rbcS genes on a pMB1 replicon were co-expressed with the groES and groEL genes on a compatible p15A replicon. Cells were grown in the presence of [^{35}S] methionine. Cell lysates were separated by SDS- and by non-denaturing polyacrylamide gel electrophoresis. The amount of soluble L_8S_8 was measured as the [^{35}S] methionine present in the Rubisco protein band in the non-denaturing gel and was normalized to the total amount of Rubisco Large subunit (LSU) as reflected by the [^{35}S]-labelled Rubisco large subunit band in the SDS-gel (*23*). a) Percent of soluble, active Rubisco relative to the total amount of LSU in the cell. b) Arrangements of co-expressed plasmids.

Table 1: The sequential order of the chaperonin assisted refolding of Rubisco. *In vitro* reconstitution of urea-denatured Rubisco was performed as specified in ref. *22*; however, [cpn60]$_{14}$, [cpn10]$_7$, Mg-ATP were applied in different orders to the unfolded Rubisco substrate

	1	2	3	4	5
Time of addition					
0'	cpn60	cpn60	cpn60	cpn60	-
	cpn10	-	-	cpn10	cpn10
	MgATP	MgATP	-	MgATP	-
20'	-	-	-	Glucose/ Hexokinase	-
23'	Rubisco	Rubisco	Rubisco	Rubisco	Rubisco
33'	-	cpn10	cpn10	-	cpn60
34'	-	-	MgATP	-	MgATP
94'	assay	Assay	assay	assay	assay
Reconstitution	45.5%	21.9%	48.3%	0.5%	0%

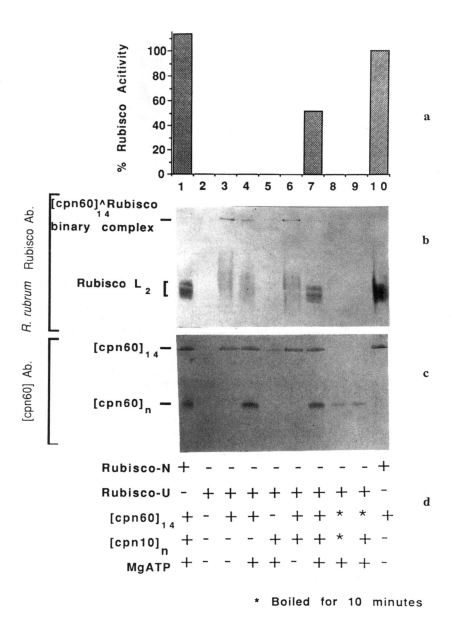

Figure 2: Chaperonin- and Mg-ATP-dependent *in vitro* reconstitution of active dimeric Rubisco from denatured Rubisco. **a**) Rubisco activity, expressed as a percentage of activity observed with an equal quantity of native Rubisco. **b**) Western blot after non-denaturing PAGE of the reaction mixtures used in a and probed with antibody raised against Rubisco. **c**) as in b, but probed with antibody raised against cpn60. **d**) Summary of the reaction conditions. Reprinted by permission from *NATURE* vol. *342* pp. 884-889. Copyright (C) 1989 Macmillan Magazines Ltd.

2B lane 3-4). The requirement for Mg-ATP in the chaperonin reconstitution reaction is shown in Table 1, lane 4, by the removal of ATP with hexokinase treatment (table 1, lane 4).

The greater the delay between the removal of the denaturant and the addition of the [cpn60]$_{14}$ complex, the more Rubisco aggregates irreversibly (22). Thus, the chaperonin prevents aggregation from occurring but is unable, in these conditions, to rescue Rubisco which has already aggregated.

ATP Hydrolysis. The chaperonin-dependent formation of active Rubisco is accompanied by the hydrolysis of ATP since non-hydrolysable analogs of ATP are ineffective substitutes. Because of technical difficulties, the stoichiometry (mols ATP hydrolyzed per mol of Rubisco refolded) remains to be determined. In the absence of an unfolded protein, cpn60 catalyses the uncoupled hydrolysis of ATP only in the presence of a monovalent cation (K^+, Rb^+ or NH_4^+ ions, but not Na^+, Li^+ or Cs^+ ions). Similarly, the chaperonin-dependent reconstitution of Rubisco shows a dependence on the same monovalent cations (4). The addition of cpn10 inhibits almost completely the uncoupled hydrolysis of ATP catalyzed by cpn60 while a [cpn60-cpn10]$_{14}$ complex is being formed (4), indicating that cpn10 functions as a coupling factor, between the hydrolysis of ATP and the protein folding reaction, bothe taking place on the [cpn60]$_{14}$ oligomer.

The Specificity Of Interaction Between cpn60 And cpn10. The *in vitro* reconstitution of dimeric Rubisco can be performed with [cpn60]$_{14}$ from different sources (Figure 3). The purification of organelle cpn60 was performed under similar conditions as for the groEL protein (24, Lorimer personal communication). Similar amounts of purified cpn60 from yeast mitochondria and pea chloroplast substituted purified *E. coli* groEL in an *in vitro* refolding assay as described in ref. 22. Although the Rubisco activities in figure 3 do not reflect the initial rates of the refolding reactions (CO_2 fixation activity was measured after one hour of refolding assay), the bacterial cpn10 is clearly able to couple the reconstitution of Rubisco in a Mg-ATP dependent manner on [cpn60]$_{14}$, despite the heterologous nature of the complex. In a reciprocal fashion, chaperonin-10 from beef liver mitochondria can functionally interact with bacterial cpn60 (24). The fact that this interaction has been conserved through evolution, from bacteria to chloroplasts and mitochondria, implies that the formation of a [cpn60-cpn10]$_{14}$ complex is important for the chaperonin function.

A Model For The Chaperonin Action On The Reconstitution Of Dimeric Rubisco. Native Rubisco dimers are unfolded by urea or partially denatured by an acid treatment. Upon removal of the denaturant, an unstable folding intermediate, Rubisco-I (Figure 4a) is thought to be formed. At low, non-physiological temperature (10°C), Rubisco-I slowly and spontaneously renatures. At 25°C however, Rubisco-I aggregates irreversibly if [cpn60]$_{14}$ is absent from the solution. In the presence of [cpn60]$_{14}$, a rather stable binary complex [cpn60]$_{14}$^Rubisco-I is formed. The *in vitro* formation of this binary complex does not require the presence of [cpn10]$_7$: However, the most probable form of the chaperone complex to interact *in vivo* with the nascent polypeptides is [cpn60-cpn10]$_{14}$ (Figure 4b).

Potassium ions are not required to form the [cpn60-cpn10]$_{14}$^Rubisco complex (4) but are required for the coupled hydrolysis of ATP and the release of the folded protein. Since at 10°C, Rubisco is able to refold without the assistance of a chaperone, we believe that the free energy of ATP is not conserved in the folded protein or the chaperones but is ultimately dissipated as heat. The manner in which

a

b

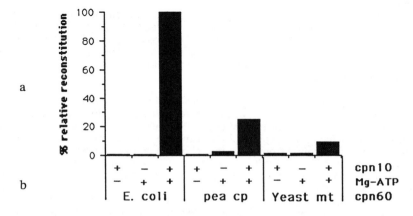

Figure 3: The ubiquitous mechanism of [cpn60]14. Purified [cpn60]14 from *E. coli*, Pea chloroplasts (pea cp) and yeast mitochondria (yeast mt) were assayed *in vitro* for their ability to reconstitute *R. rubrum* Rubisco in the presence (+) or absence (-) of *E. coli* cpn10 and Mg-ATP as in ref (22). **a**) Percent reconstitution after one hour assay, measured as in Figure 2a, normalized to the *E. coli* cpn60 standard reconstitution reaction. **b**) Summary of the reaction conditions.

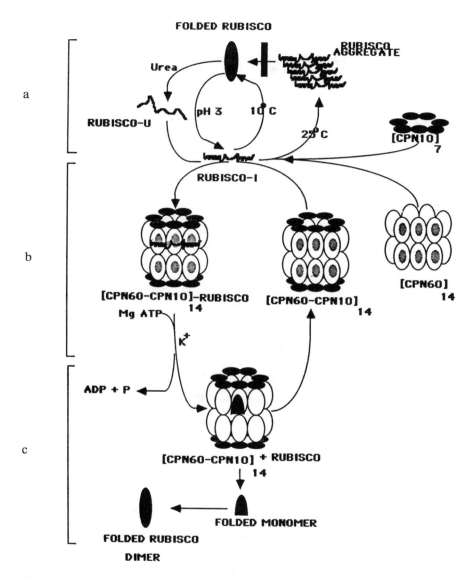

Figure 4: Model for chaperonin-dependent reconstitution of Rubisco. Native Rubisco (Rubisco-N) is denatured by urea, guanidium–HCl, or acid (pH 3.0), to form after removal of the denaturing agent; Rubisco-I (a) an unstable folding intermediate. Rubisco-I aggregates, unless (b) the chaperonin $[cpn60]_{14}$ or $[cpn60–cpn10]_{14}$ associate with it. The complex $[cpn60–cpn10]_{14}$-Rubisco requires Mg–ATP and K^+ ions to catalyze the refolding reaction (c) and the formation of a monomeric Rubisco-N that spontaneously dimerizes to form active Rubisco.

this energy is utilized during chaperonin-facilitated folding processes remains to be determined.

Conclusion. We have provided experimental evidence that the chaperonins meet the first criterion expected by their definition. They are molecules that, by their transient association with nascent, stress-destabilized or translocated proteins, prevent "improper" aggregation. Chaperonins are not able to rescue aggregates once they are formed, nor do they appear to carry specific steric information, capable of directing a protein to assume a structure different from the one dictated by polypeptide primary sequence.

Literature Cited.

1   Laskey, R. A., Honda, B. M., Mills, A. D. and Finch, J. T. *Nature* **1978**, *275*, 416-420.
2   Barraclough, R. and Ellis R. *J. Biochim. Biopys. Acta* **1980**, *608*, 19-31.
3   Anfinsen, C. B. *Science* **1973**, *181*, 223-230.
4   Viitanen, P. V., Lubben, T. H., Reed, J., Goloubinoff, P., O'Keefe, D. P. and Lorimer, G. H. *Biochemistry* **1990**, *29*, 5665-5671.
5   Pelham, H. R. B.*Cell* **1986**, *46*, 959-961.
6   Ellis R. J., van der Vies, S. M. and Hemmingsen, S. M. *Biochem. Soc. Symp.* **1989**, *55*, 145-153.
7   Ellis, R. J. *Nature* **1987**, *328*, 378-379.
8   Ellis, R. J. and van der Vies, S. M.*Photosynthesis Res.* **1988**, *16*, 101-115.
9   Roy, H. and Cannon, S.*Trends Biochem. Sci.* **1988**, *13*, 163-165.
10   Hemmingsen, S. M., Woolford, C., van der Vies, S. M., Tilly, K., Dennis, D. T., Georgopoulos, C. P., Hendrix, R. W. and Ellis, R. J. *Nature* **1988**, *333*, 330-334.
11   Georgopoulos, C. P. and Hohn, B. *Proc. Natl. Acad. Sci. U.S.A.* **1978**, *75*, 131-135.
12   Zweig, M. and Cummings, D. J. *J. Molec. Biol.* **1973**, *80*, 505-518.
13   Tilly, K., Murialdo, H. Georgopoulos, C. P. *Proc. Natl. Acad. Sci. U.S.A.* **1981**, *78*, 1629-1633.
14   Kochan, J. and Murialdo, H. *Virology* **1983**, *131*, 100-115.
15   Gatenby, A. A., van der Vies, S. M. and Bradley, D. *Nature* **1985**, *314*, 617-620.
16   Goloubinoff, P., Gatenby, A. A. and Lorimer, G. H. *Nature* **1989**, *337*, 44-47.
17   Fayet, O., Louarn, J-M. and Georgopoulos, C. P. *Molec. Gen. Genet.* **1986**, *202*, 435-445.
18   Jenkins, A. J., March, J. B., Oliver, I. R. and Masters, M. *Molec. Gen. Genet.* **1986**, *202*, 446-454.
19   Van Dyk, T. K., Gatenby, A. A. and LaRossa, R. A. Nature 1989, 342, 451-453.
20   Cheng, M. Y., Hartl, F.-U., Martin, J., Pollock, R. A., Kalousek, F., Neupert, W., Hallberg, E. M., Hallberg, R. L. and Horwich, A. L. *Nature* **1989**, *337*, 620-625.
21   Lubben, T. H., Donaldson, G. K., Viitanen, P. V. and Gatenby, A. A. *Plant Cell.* **1989**, *1*, 1223-1230.
22   Goloubinoff, P., Christeller, J. T., Gatenby, A. A. and Lorimer, G. H. *Nature* **1989**, *342*, 884-889.
23   Laemmli, U. K. *Nature New Biol.* **1970**, *277*, 680-685.
24   Lubben, T. H., Gatenby, A. A., Donaldson,G. K., Lorimer, G. H. and Viitanen, P. V. *Proc. Natl. Acad. Sci. U.S.A.* **1990**, *in press.*

RECEIVED February 6, 1991

Chapter 10

Pathway for the Thermal Unfolding of Wild Type and Mutant Forms of the Thermostable P22 Tailspike Endorhamnosidase

Bao-lu Chen and Jonathan King

Department of Biology, Massachusetts Institute of Technology, Cambridge, MA 02139

Increasing the stabilities of proteins under realistic conditions requires understanding the mechanisms of protein denaturation far from equilibrium. A protein displaying very high thermal stability, resistance to proteases and resistance to detergents is the tailspike endorhamnosidase of bacteriophage P22. The folding pathway for the tailspike is well defined both *in vivo* and *in vitro*. Kinetic analysis of the thermal and detergent unfolding pathway reveals a relatively slow process which passes through a long-lived partially folded trimeric intermediate. This species has its N-termini unfolded with the remaining regions of the polypeptide chains in a compact native-like form. The intermediate can refold back to the native on cooling. Further unfolding of the intermediate at high temperature generates an aggregating species which renders the process kinetically irreversible. In the presence of SDS the unfolding intermediate is quantitatively converted to an unfolded monomer/SDS complex. Thus denaturation initiates via the melting of sites at the N-terminus, but the rate limiting step for the overall process is the melting of the intermediate species. Examining the thermal unfolding kinetics of more than ten temperature sensitive for folding (*tsf*) mutant proteins shows they have small effects on the transition from the native to the intermediate, but large effects on the transition from the intermediate to the fully unfolded form. The genetic modification of unfolding intermediates may provide an avenue for increasing the stability of proteins to irreversible denaturation.

One goal of protein engineering is to increase the stability of proteins of value by substituting amino acids at sites affecting stability (*1*). Investigations of protein stability have generally focused on small globular proteins under conditions in which equilibrium of a reversible unfolding/refolding transition can be achieved

0097–6156/91/0470–0119$06.00/0

(2,3). Obtaining such behavior usually requires denaturants such as urea or GdnHCl. The unfolding transitions observed under these conditions are generally highly cooperative with intermediate states not significantly populated (4,5). At equilibrium protein stability as defined by the free energy change between the native and unfolded states can be estimated by means of the two-state model (6). By extrapolating to zero denaturant (7,8) the stability determined from urea or GdnHCl denaturation measurements fits quite well with that obtained from thermal unfolding (9). Stability exhibits a maximum at some temperature and declines at both higher and lower temperatures (6,10).

The combination of molecular modeling with genetic engineering to enhance protein stability has been successful in certain cases. For instance, introducing carefully sited novel disulfide bonds increased protein stability in T4 lysozyme (11-13) and in λ-repressor (14). However, the results in other proteins, for instance, in subtilisin (15,16) and in dihydrofolate reductase (17) have been less predictable.

Attempts have been made to interpret the effects of single and multiple mutant proteins on thermodynamically defined stability at a structural level (18,19). The majority of these results have been interpreted in terms of the substitutions destabilizing the native structure (19,20). In contrast Shortle and Meeker (21) proposed that some mutations act by altering the stability of the unfolded state. These approaches have not generally led to an understanding of the actual pathway of denaturation or of the chemical nature of the rate limiting step in the reaction.

Conditions used for carrying out reversible unfolding equilibrium studies are quite different from the conditions in which thermostability is biologically or economically relevant, such as in detergent solution in a washing machine, under fermentation conditions, in the intestinal tract or in the bloodstream.

Denaturation at high temperature and other realistic conditions is rarely reversible (22,23). Depending on conditions a component of the irreversibility may be due to covalent damage to the polypeptide chain (23). However a major source of irreversibility is the conformational transition to an aggregated state in which the chains are kinetically trapped (24). In both the forward and reverse pathways the aggregation step derives from a property of the conformation of the folding/unfolding intermediates (24, Mitraki, A. and King, J., this volume).

For these multi-step folding/unfolding processes complicated by aggregation, calculating the stability of a protein via the two-state model is not valid because the denatured state is not in equilibrium with the native state. An alternative approach, kinetic analysis of the pathway of unfolding, is required to study protein thermostability. The stability of intermediate species along the unfolding pathway and the energy barriers of the unfolding play important roles during unfolding.

Mutations that influence the folding/unfolding pathway *in vitro* through their effects on intermediates have been described by Matthews and coworkers for the α-subunit of tryptophan synthetase (25). A large set of mutations affecting an intermediate in the *in vivo* folding pathway have been characterized in the P22 tailspike (26-29). Mutants which affect the unfolding pathway by affecting the stability of the intermediates, or the energy barrier (the transition state) may have little effect on the overall stability of the native protein. Characterization of these mutations requires examination of their kinetic effects on the unfolding transition.

The tailspike protein from bacteriophage P22 is well suited for the second approach. The tailspike is a structural protein of P22. It is the last protein to bind to virus capsids during morphogenesis. The tailspike is also an endorhamnosidase which cleaves the O-antigen protruding from its host cell *Salmonella* upon infection (*30*). Its secondary structure is dominated by β-sheets probably in a cross-β conformation. This mesophilic enzyme is highly thermostable and has an apparent melting temperature of 88°C as measured by both microcalorimetry and Raman spectroscopy (*31-33*). Such high thermostability could either result from specific structural features in the native protein or from energy barriers on the unfolding pathways.

During the forward folding pathway the newly synthesized chain forms a partially folded single chain intermediate (*34,35*). This folds further to a species with sufficient structure for chain recognition. Three of these intermediates associate into a protrimer in which the chains are associated but not fully folded (*36*). This species then folds further to the thermostable detergent and protease resistant native trimer. There are no known covalent modifications in the process. A similar process occurs in refolding urea denatured chains *in vitro* (*37*).

Thermal Unfolding Pathway of Tailspike

The tailspike protein is composed of three identical polypeptide chains of 666 residues (*27,38*). The native trimer is resistant to SDS denaturation as well as to proteolytic digestion, in addition to its high stability to heat denaturation. Since the protein remains native in SDS solutions at room temperatures, it binds few SDS molecules and migrates anomalously slowly in SDS gels (*27*). By contrast, the SDS/polypeptide chain complex formed from the denatured chain (*39*) migrates proportionally to its molecular weight during SDS gel electrophoresis (*27*). We have taken advantage of these aspects of SDS-PAGE to follow the thermal unfolding pathway (*40*).

Thermal Unfolding of Tailspike via a Sequential Pathway.   The thermal unfolding of the tailspike protein was examined by first incubating the protein samples prepared in Tris buffer near neutral pH at high temperature for various amounts of time and then examining the thermal unfolding products by SDS-PAGE at room temperature (about 20°C). The unfolding rate, measured as the loss of the species of native mobility with time, was quite slow even at temperatures as high as 80°C. The unfolding pathway proceeded through a discrete intermediate, whose SDS electrophoretic mobility was between that of the native tailspike and that of the fully denatured tailspike/SDS complex.   This intermediate species "*I*" was produced upon heating the native trimer "*N*", but appeared before the fully unfolded monomers "*M*". It electrophoresed slightly faster than the native trimer but slower than the unfolded monomer thus allowing detection.

The intermediate could be trapped by chilling and was stable in the cold for greater than 24 hr. Upon incubation at room temperature it refolded, though quite slowly, to a species with the native mobility.

Analysis of the products of the thermal unfolding reaction revealed that after the appearance of the unfolding intermediate, recovery of the chains decreased due to aggregation. This aggregation probably resulted from the partially unfolded monomeric chains formed from the dissociation of the intermediates. The aggregation of the partially denatured chains rendered the kinetic analysis difficult.

In the presence of SDS, the unfolding rate accelerates and can be measured at lower temperature. SDS inhibits the refolding of the *I* species back to the native. There is no detectable aggregation in SDS and the transformation from *N* to *I* and then to *M* is quantitative. The unfolding mechanism of tailspike protein can be described as:

$$N \text{ (native trimer)} \xrightarrow{k_1} I \text{ (trimeric intermediate)} \xrightarrow{k_2} M \text{ (unfolded monomers)}$$

Here, k_1 and k_2 are the unfolding rate constants for the unfolding transition from N to I and from I to M, respectively. The unfolding kinetics can be quantitatively analyzed from the scanned intensities of the Coomassie blue stained tailspike bands on SDS gels. Figure 1 depicts results from a typical thermal unfolding experiment for wild type tailspike protein, which was performed in Tris buffer (pH 8) and 2% SDS at 65°C. Kinetic analysis yields two rate constants: $1.1 \times 10^{-3} \text{ s}^{-1}$ and $4.0 \times 10^{-5} \text{ s}^{-1}$ for the conversion from N to I and from I to M, respectively.

The Thermal Unfolding Rate Constants Strongly Depend on pH. Investigation of the pH dependence of thermal unfolding showed that both k_1 and k_2 depend strongly on the solvent pH. This was examined in 2% SDS at 65°C at which both thermal unfolding rate constants are relatively easy to measure. The results are presented in plots of $\log k_1$ and $\log k_2$ versus pH in Figure 2. Both rate constants showed biphasic behaviors: reaching a minimum near neutral pH and increasing at lower pH as well as at higher pH. Thus the unfolding reaction is both acid and base catalyzed. The native protein is most stable at pH 7 and the intermediate at pH 6.5. The slopes of the $\log k_1$ versus pH curve for the low pH phase and high pH phase are -0.46 and 0.35, respectively. Correspondingly, the slopes of $\log k_2$ versus pH curve for the low pH phase and high pH phase are -1.9 and 1.1, respectively. If these slopes measure the ionization of particular charge groups which participate in the formation of the transition states of unfolding (22,41), more charged groups must be involved in the rate determining step of the transition from I to M than from N to I.

The Thermal Unfolding Intermediate Is Partially Unfolded at its N-terminus. The mobility of the intermediate in the SDS gel as well as its sedimentation behavior in a sucrose gradient indicated that I is still trimeric (40). The slight increase in the gel electrophoretic mobility presumably resulted from increased binding of SDS molecules due to local unfolding. The partial unfolding also makes this species sediment slightly slower than the native trimer.

The tailspike has eight buried cysteines spanning the polypeptide chain from residues 169 to 635 (38). In the presence of SDS, all of the cysteines remained insensitive to Ellman's reagent both at the intermediate stage and in the native conformation (40). All of them were reactive when the chains were fully unfolded. In the absence of SDS, trypsin cleaved the intermediate species progressively from the N-termini. The cleavage stopped at Lys-107 and left a species corresponding to about 550 amino acid polypeptide associated chains (40). Probably the intermediate species generated at elevated temperatures retained minor residual structures at its N-termini which slowed down the trypsin cleavage. Nevertheless, results from both cysteine exposure measurement and protease cleavage of the intermediate species suggest the unfolding at the intermediate stage is limited to the N-terminal region.

The purified tailspikes can bind to phage capsids *in vitro* and make the resulting phage particles infectious (42). The thermal unfolding intermediate was unable to bind to the phage capsid (40). The specific binding activity to phage

Time of Incubation at 65°C (min)

3 6 10 15 30 50 80 120 200 0

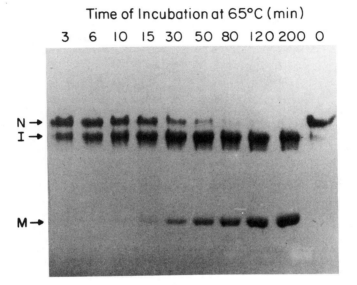

Figure 1. Thermal unfolding of wild type tailspike protein at 65°C. Thermal unfolding was performed by incubating 0.4 mg/ml tailspike prepared in 50 mM Tris (pH 8), 1.7 mM 2-mercaptoethanol and 2% SDS at 65°C. Samples were taken at the indicated times. The reaction was quenched by mixing the samples with SDS sample buffer (62.5 mM Tris at pH 7, 2.1 mM 2-mercaptoethanol, 10% glycerol, 0.012% Bromophenol blue dye and 2% SDS) in the cold. Then the samples were electrophoresed through SDS-PAGE at about 20°C and the proteins were stained with Coomassie blue.

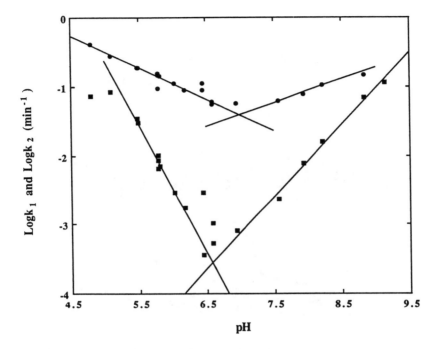

Figure 2. Dependence of unfolding rate constants on pH. Wild type tailspike protein was prepared in 50 mM Tris, 1.7 mM 2-mercaptoethanol and 2% SDS and adjusted to different pH values by 1 N HCl. Thermal unfolding was done at 65°C and followed by SDS-PAGE at about 20°C. Sample pH values shown here have been corrected to 65°C. k_1 (●) and k_2 (■) shown in log are the thermal unfolding rate constants for the conversions from N to I and from I to M, respectively. The linear lines through the data points are the results of least-square fit to each individual pH phase for both k_1 and k_2 data. The calculated slopes of the fitting lines for k_1 are -0.46 and 0.35 for the low and high pH phases, respectively; and for k_2 are -1.9 and 1.1 for the low and high pH phases, respectively.

capsids decreased corresponding to the increase of the concentration of the intermediate in the reaction mixture in the presence of SDS. Thus partial unfolding at the N-termini abolishes its binding activity to the phage capsid.

Thermal Unfolding Is not Melting of Two Independent Domains. The native tailspike protein shows symmetrical folding of the three polypeptide chains, as concluded from low resolution X-ray crystallographic results (Alber, T., University of Utah, personal communication). Each subunit probably extends the full length of the tailspike so that the contact interfaces between the subunits would be quite large. Given the estimated 50% - 60% β-strand content (*31*), β-sheets might play an important role in the formation of the subunit interfaces. Both the trimeric feature of the thermal unfolding intermediate from our analysis and the single thermal unfolding transition peak measured by microcalorimetry suggest that the interaction between subunit interfaces is extremely strong.

Sucrose gradient sedimentation of an N-terminal 55-kDa amber tailspike polypeptide fragment shows that it is still in a monomeric form (*43*). Thus the missing C-terminal region is critical for trimer formation. These results suggest that the N-terminal 110 amino acids are apparently not able to form an independent folding domain in the native conformation, even they are denatured at the first stage of unfolding.

Thermal Unfolding Pathway Differs from the *in vivo* Folding Pathway

The intracellular folding mechanism of the P22 tailspike has been extensively characterized by genetic analysis (Figure 3). The maturation process of tailspike protein inside the cell proceeds through several defined intermediate stages. Three newly synthesized polypeptide chains first fold into conformations which are ready for association and then fold/associate to a trimeric intermediate, the protrimer, and finally fold into the native trimer (*34,44*). The half time for the monomeric chains to fold into a SDS- resistant native trimer *in vivo* at 30°C is about 5 min (*36*).

Comparison of Thermal Unfolding Intermediate to Protrimer. Upon examination of the forward *in vivo* folding products by SDS-PAGE, no species was found corresponding to the thermal unfolding intermediate. The late folding intermediate is the protrimer which can be identified by non-denaturing gel electrophoresis and has partially folded and associated structure (*36*). It can be trapped in the cold and will subsequently fold to the native state after shifting up the temperature. In the presence of SDS, the protrimer is not stable and unfolds to the denatured monomeric form. The protrimer is also sensitive to trypsin digestion. Thus the *in vivo* protrimer folding intermediate is less stable than the thermal unfolding intermediate with regard to denaturation by SDS and sensitivity to proteases. These comparisons show that they are different from each other (Table I). Since these two pathways are populated by different intermediates, we conclude that the *in vitro* thermal unfolding pathway at elevated temperatures is not the simple reverse of the *in vivo* folding pathway at physiological temperatures.

If environmental variations will change the folding/unfolding mechanism, unfolding by different means might follow different pathways. For instance, the molten globule state as a protein folding intermediate is usually observed under acidic conditions (*45*). We have examined the acid unfolding pathway of tailspike protein, and found that it followed the same pathway as thermal unfolding (Chen,

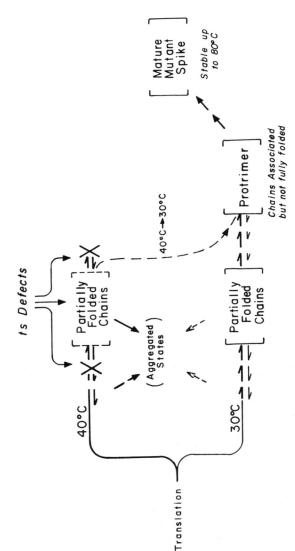

Figure 3. Intracellular folding pathway of P22 tailspike proteins. The newly synthesized wild type or mutant polypeptide chains at 30°C first fold into partially folded monomeric intermediates. These species fold and associate to form a protrimer intermediate. Further folding results in a thermostable native tailspike. At 40°C, the folding is inhibited and *tsf* mutants act by blocking an early step in chain folding, prior to association. However, if infected cells are shifted to 30°C, the mutant chains continue through the productive pathway.

Table I. Comparison between the Folding Intermediate (Protrimer) and the Thermal Unfolding Intermediate (*I*)

Property	Protrimer (*in vivo*)	*I* (*in vitro*)
conformational state	partially folded trimer	trimer with unfolded N-termini
folding to native	yes	yes
SDS resistance	no	yes
Protease resistance	sensitive	central and C-terminal regions resistant

B.-L. and King, J., unpublished results). Tailspike also unfolds in cold SDS solution (4°C) and converts to an intermediate species similar to the thermal unfolding intermediate with a half time of 10 days. However the mechanism of unfolding induced by urea and GdnHCl are still unclear and currently under investigation.

Recent reports on facilitation of protein folding by molecular chaperones (46-48) provide another possibility: the difference between these two pathways could be a consequence of lack of certain cellular components in the *in vitro* reaction. It was observed that the products of *groE*-bearing plasmid were able to rescue some temperature sensitive P22 tailspike mutants in *Salmonella* at the restrictive temperature (39°C), though very weakly (49). On the other hand, high yield of the native protein was also reported in the refolding (reconstitution) of the acid urea denatured tailspike polypeptide chains without the addition of cellular factors at low temperatures (10°C) (37). This *in vitro* result indicates that auxiliary factors inside the cell are not absolutely required for the folding of this protein, at least under these experimental conditions. However, this does not rule out that the cellular factors could play a role under *in vivo* conditions.

Unfolding of water soluble oligomeric proteins usually follows distinct stages: first subunit-subunit interactions are disrupted and then the structural monomer unfolds. In contrast, both *in vivo* folding and *in vitro* thermal unfolding of tailspike show no evidence of structural ("native") monomers. Thus the coupling between tertiary and quaternary structures for this protein is quite strong. The dominant secondary structure of this protein is β-sheet as estimated from Raman spectroscopic measurements (31). During association of the tailspike subunits particular structural features might be formed such as interchain secondary structures or interwinding of the subunits. These structures may make significant contributions to the stability of this protein.

In any case, the difference between the *in vivo* folding pathway and the *in vitro* thermal unfolding pathway indicates the importance of kinetic aspects in controlling the folding and unfolding of this protein.

Effects of Tailspike Mutations on Both Folding and Unfolding Pathways

Initiation Step of Unfolding Involves Breaking a Salt Bridge. Schwarz and Berget (*50*) isolated a mutation at the N-terminal region of the tailspike polypeptide chain, Asp100->Asn, which was defective in capsid binding activity. Mutations suppressing this defect were subsequently isolated and found to be substitutions for Arg-13 (*51*). They proposed that the breakage of a salt bridge between Arg-13 and Asp-100 was the cause of the Asp100->Asn defect.

The breakage of the salt bridge between residues 13 and 100 also rendered the N-termini of the polypeptide chains sensitive to SDS denaturation. The Asp100->Asn protein migrates similarly to the thermal unfolding intermediate during SDS electrophoresis, while it migrates as the wild type native protein during non-denaturing electrophoresis. This mutant protein is probably susceptible to partial denaturation in cold SDS solution. Substitution of Arg-13 by Ser, His and Leu gave similar results (*51*). The salt bridge formed between Arg-13 and Asp-100 must play an important role in stabilizing the N-termini of the tailspike. It is known that salt bridges are one of the contributing factors to protein thermostability (*52*). Therefore, the simplest interpretation of these results could be that the initiation step of unfolding of tailspike is the breakage of this salt bridge.

The salt bridge between Arg-13 and Asp-100 revealed from Berget's work may well explain the pH dependence of the unfolding rate constant k_1. Presumably, the breakage of this salt bridge is critical in the rate determining step of the conversion from N to I. The two phases for the pH dependence of k_1 could be attributed to the ionization of Asp-100 and Arg-13. Probably, approximate 50% protonation of the Asp-100 group or deprotonation of the Arg-13 group will be sufficient to allow the conversion to take place.

Detailed analysis of the pH dependence of the unfolding rate constant k_2 has not yet been completed. Still unknown is which charge groups are involved in the rate determining step for the conversion from I to M. SDS molecules carry negative charges which will certainly affect the protonation of the acidic groups. This is probably one of the reasons why the slopes of the low pH phase for both k_1 and k_2 are larger than those of the corresponding high pH phase.

Ts Tailspikes Block Folding at the Restrictive Temperature. Temperature sensitive mutations have been found in the gene coding for tailspike protein in P22 phage. Most have a single amino acid residue replacement and have been mapped to the central region of the chain (*53,54*). These are temperature sensitive for folding (*tsf*) mutations. They block folding at restrictive conditions but fold normally at permissive conditions. The "permissively" matured *tsf* proteins function normally at restrictive temperatures. Microcalorimetric and Raman spectroscopic measurements of a number of *tsf* mutants showed they have almost identical melting temperatures as wild type. This melting temperature is about 40°C above the restrictive temperature at which the folding defect occurs (*32,33*). These mutants identify critical sites in the polypeptide chain which determine the correct folding pathway.

Tsf Tailspikes Affect the Second Step of Thermal Unfolding. A surprising aspect of the *tsf* substitutions was their very limited effect on the apparent melting temperature of thermal denaturation (*32,33*). The *tsf* mutant proteins could affect later steps in the unfolding pathway. We have performed kinetic analysis of the thermal unfolding for more than ten *tsf* mutant proteins and the detailed results of this study will be reported elsewhere (Chen, B.-L., Yu, M.-H., and King, J., unpublished results). In those experiments, thermal unfolding was carried out in 2% SDS, Tris buffer near neutral pH at temperatures between 60°C and 70°C. All of the mutant proteins follow an unfolding pathway similar to the wild type. With one of the mutant proteins, tsU38 (Gly435->Glu), the thermal unfolding intermediate is largely destabilized and becomes a transient species. Examined by Raman spectroscopic measurements, this mutant protein is known to melt at a lower temperature as compared to the wild type protein (*31*) while other *tsf* mutant tailspikes melt similarly as the wild type protein (*33*).

Kinetic analysis of the thermal unfolding of these mutant proteins shows that they all have small effects on k_1, the rate constant for the transition from the native to the intermediate, but large effects on k_2, the rate constant for the transition from the intermediate to the fully unfolded monomers. For example, Figure 4 displays the thermal unfolding kinetics for both wild type and mutant tsU57 (Asp230->Val) proteins at 70°C. The mutant protein tsU57 had unfolding rate constants $k_1 = 4.3 \times 10^{-3}$ s^{-1} and $k_2 = 5.8 \times 10^{-4}$ s^{-1}, while the wild type protein had unfolding rate constants $k_1 = 4.0 \times 10^{-3}$ s^{-1} and $k_2 = 5.5 \times 10^{-5}$ s^{-1} under these conditions. The single amino acid substitution in this particular instance slightly alters the unfolding of the native state, increasing the unfolding rate 1.1-fold with respect to the wild type. But this mutation does destabilize the intermediate species which melts 11-fold faster than that of the wild type. These results demonstrate that unlike *ts* mutations found in T4 lysozyme which mark the labile sites of the native state (*55,56*), *tsf* mutations in P22 tailspike identify labile sites of the intermediates.

The small effect of these *tsf* substitutions on the first kinetic phase for thermal unfolding explains why they have hardly any effect on the melting temperature. Unfolding of the N-terminal region of the molecule is the initiation step for thermal denaturation of this protein. The apparent Tm measured by microcalorimetry could be defined by this step. Given their location in the central region of the chain, it is not surprising that the *tsf* mutant proteins affect the second step but not the first.

Tailspike *tsf* mutants were originally isolated because they destabilize the early *in vivo* folding intermediates. The present work shows that they also have a large kinetic effect on the second transition of the thermal unfolding pathway. Some common properties may be shared in these two pathways. For instance, the intermediates found in the early folding stage might have similar structural features to the intermediates present in the late stage of unfolding. Further work has to be undertaken to answer this question.

Conclusions

Single residue substitution in a protein can have a dramatic effect on its overall thermal stability. For example, a 17°C change in Tm was observed for an Asn57 ->Ile replacement in yeast iso-1-cytochrome c (*57*). Single residue substitution in a protein can also cause a large variation on the kinetics of unfolding. For example, the above mentioned mutant protein tsU57 has 1.1-fold and 11-fold change for the unfolding rate constants, k_1, and k_2, respectively, as compared to the wild type protein. Thus in the effort to engineer proteins against heat

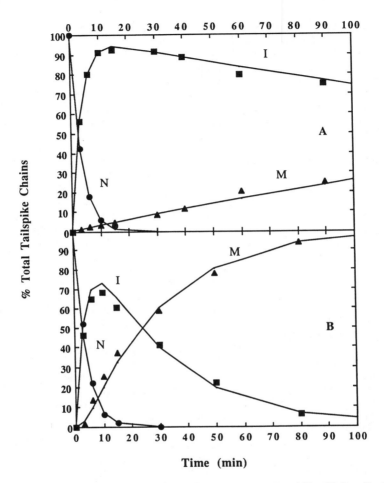

Figure 4. Thermal unfolding of wild type and tsU57 (Asp230->Val) tailspike proteins at 70°C. Tailspike proteins in 50 mM Tris (pH 6.8), 0.71 M 2-mercaptoethanol and 2% SDS were subjected to thermal unfolding at 70°C followed by SDS-PAGE at about 20°C. The scanned intensities of Coomassie blue stained tailspike N (●), I (■) and M (▲) bands on SDS gels were plotted as the percentage of total tailspike protein represented by each species for both wild type (panel A) and tsU57 (panel B). The lines are the fitting results according to the unfolding mechanism: $N \rightarrow I \rightarrow M$ with k_1 as rate constant for $N \rightarrow I$ and k_2 as rate constant for $I \rightarrow M$. Under this condition, k_1 for wild type and tsU57 is 4.0×10^{-3} s^{-1} and 4.3×10^{-3} s^{-1}, respectively; k_2 for wild type and tsU57 is 5.5×10^{-5} s^{-1} and 5.8×10^{-4} s^{-1}, respectively.

denaturation, in addition to searching for sites which contribute to overall thermostability of the native state, we might also search for sites which are crucial for the stability of the intermediates and unfolding barriers. That is because protein thermostability depends not only on the coordination of forces in the native protein but also on the stability of the intermediate and the barrier of the transition state in the early unfolding stage.

Acknowledgments

This work was supported by NIH Grant GM17980 and NSF grant under the Engineering Research Center Initiative to the Biotechnology Process Engineering Center (Cooperative Agreement ECD88-03014).

Literature Cited

1. Matthews, B. W. *Biochemistry* **1987**, *26*, 6885.
2. Privalov, P. L. *Adv. Protein Chem.* **1979**, *33*, 167.
3. Schellman, J. A. *Ann. Rev. Biophys. Biophys. Chem.* **1987**, *16*, 115.
4. Privalov, P. L.; Khechinashvili, N. N. *J. Mol. Biol.* **1974**, *86*, 665.
5. Kim, P. S.; Baldwin, R. L. *Ann. Rev. Biochem.* **1982**, *51*, 459.
6. Becktel, W. J.; Schellman, J. A. *Biopolymers* **1987**, *26*, 1859.
7. Schellman, J. A. *Biopolymers* **1978**, *17*, 1305.
8. Pace, N. C. *Methods Enzymol.* **1986**, *131*, 266-280.
9. Pace, N. C. *Trends in Biochem. Sci.* **1990**, *15*, 14.
10. Chen, B.-L.; Schellman, J. A. *Biochemistry* **1989**, *28*, 685.
11. Wetzel, R.; Perry, L. J.; Baase, W. A.; Becktel, W. J. *Proc. Natl. Acad. Sci. U.S.A.* **1988**, *85*, 401.
12. Matsumura, M.; Becktel, W. J.; Levitt, M.; Matthews, B. W. *Proc. Natl. Acad. Sci. U.S.A.* **1989**, *86*, 6562.
13. Matsumura, M.; Signor, G.; Matthews, B. W. *Nature* **1989**, *342*, 291.
14. Sauer, R. T.; Hehir, K.; Stearman, R. S.; Weiss, M. A.; Jeitler-Nilsson, A.; Suchanek, E. G.; Pabo, C. O. *Biochemistry* **1986**, *25*, 5992.
15. Wells, J. A.; Powers, D. B. *J. Biol. Chem.* **1986**, *261*, 6564.
16. Pantoliano, M. W.; Ladner, R. C.; Bryan, P. N.; Rollence, M. L.; Wood, J. F.; Poulos, T. L. *Biochemistry* **1987**, *26*, 2077.
17. Villafranca, J. E.; Howell, E. E.; Oatley, S. J.; Xuong, N.-H.; Kraut, J. *Biochemistry* **1987**, *26*, 2182.
18. Goldenberg, D. P. *Ann. Rev. Biophys. Biophys. Chem.* **1988**, *17*, 481-507.
19. Alber, T. *Ann. Rev. Biochem.* **1989**, *58*, 765.
20. Wells, J. A. *Biochemistry* **1990**, *29*, 8509.
21. Shortle, D.; Meeker, A. K. *Proteins* **1986**, *1*, 81.
22. Tanford, C. *Adv. Protein Chem.* **1968**, *23*, 121.
23. Volkin, D. B.; Klibanov, A. M. In *Protein Function: a Practical Approach;* Creighton, T. E., Eds.; IRL Press, Oxford, 1989, pp. 1-24.
24. Mitraki, A.; King, J. *Bio/Technology* **1989**, *7*, 690.
25. Beasty, A. M.; Hurle, M. R.; Manz, J. T.; Stackhouse, T.; Onuffer, J. J.; Matthews, C. R. *Biochemistry* **1986**, *25*, 2965.
26. Smith, D. H.; Berget, P. B.; King, J. *Genetics* **1980**, *96*, 331.
27. Goldenberg, D. P.; Berget, P. B.; King, J. *J. Biol. Chem.* **1982**, *257*, 7864.
28. Yu, M.-H.; King, J. *Proc. Natl. Acad. Sci. U.S.A.* **1984**, *81*, 6584.

29. King, J; Fane, B.; Haase-Pettingell, C.; Mitraki, A.; Villafane, R.; Yu, M.-H. In *Protein Folding: Deciphering the Second Half of the Genetic Code;* Gierasch, L.; King, J. Eds.; AAAS, Washington D. C. 1989, pp. 225-239.
30. Iwashita, S.; Kanegasaki, S. *Eur. J. Biochem.* **1976**, *65*, 87.
31. Sargent, O.; Benevides, J. M.; Yu, M.-H.; King, J; Thomas, G. J., Jr. *J. Mol. Biol.* **1988**, *199*, 491.
32. Sturtevant, J. M.; Yu, M.-H.; Haase-Pettingell, C.; King, J. *J. Biol. Chem.* **1989**, *264*, 10693.
33. Thomas, G. J.; Becka, R.; Sargent, D.; Yu, M.-H.; King, J. *Biochemistry* **1990**, *29*, 4181.
34. Goldenberg, D. P.; Smith, D. H.; King, J. *Biopolymers* **1983**, *22*, 125.
35. Haase-Pettingell, C.; King, J. *J. Mol. Chem.* **1988**, *263*, 4977.
36. Goldenberg, D.; King, J. *Proc. Natl. Acad. Sci. U.S.A.* **1982**, *79*, 3403.
37. Seckler, R.; Fuchs, A.; King, J.; Jaenicke, R. *J. Biol. Chem.* **1989**, *264*, 11750.
38. Sauer, R. T.; Krovatin, W.; Poteete, A. R.; Berget, P. *Biochemistry* **1982**, *21*, 5811.
39. Reynolds, J.; Tanford, C. *J. Biol. Chem.* **1970**, *245*, 5161.
40. Chen, B.-L.; King, J. *Biochemistry* **1991**, submitted.
41. Steinhardt, J.; Zaiser, E. M. *J. Am. Chem. Soc.* **1953**, *75*, 1599.
42. Israel, J. V.; Anderson, T. F.; Levine, M. *Proc. Natl. Acad. Sci. U.S.A.* **1967**, *57*, 284.
43. Friguet, B.; Djavadi-Ohaniance, L.; Haase-Pettingell, C. A.; King, J.; Goldberg, M. E. *J. Biol. Chem.* **1990**, *265*, 10347.
44. Goldenberg, D.; Smith, D.; King, J. *Proc. Natl. Acad. Sci. U.S.A.* **1983**, *80*, 7060.
45. Ptitsyn, O. B. *J. Prot. Chem.* **1987**, *6*, 273.
46. Rothman, J. E. *Cell* **1989**, *59*, 591.
47. Schlesinger, M. J. *J. Biol. Chem.* **1990**, *265*, 12111.
48. Ellis, R. J. *Science* **1990**, *250*, 954.
49. Van Dyk, T. K.; Gatenby, A. A.; LaRossa, R. A. *Nature* **1989**, *342*, 451.
50. Schwarz, J.; Berget, P. B. *J. Biol. Chem.* **1989**, *264*, 20112.
51. Maurides, P.; Schwarz, J. J.; Berget, P. B. *Genetics* **1990**, *125*, 673.
52. Perutz, M. F. *Science* **1978**, *201*, 1187.
53. Smith, D. H.; King, J. *J. Mol. Biol.* **1981**, *145*, 653.
54. Villafane, R.; King, J. *J. Mol. Biol.* **1988**, *204*, 607.
55. Hawkes, R.; Grutter, M. G.; Schellman, J. A. *J. Mol. Biol.* **1984**, *175*, 195.
56. Grutter, M. G.; Hawkes, R. B.; Matthews, B. W. *Nature* **1979**, *277*, 667.
57. Das, G; Hickey, D. R.; Mclendon, D; Sherman, F. *Proc. Natl. Acad. Sci. U.S.A.* **1989**, *86*, 496.

RECEIVED March 8, 1991

Chapter 11

Kinetics of Inclusion Body Formation in *Escherichia coli* CY15070 Producing Prochymosin

Alfred Carlson and Jayanthi Reddy

Department of Chemical Engineering, The Pennsylvania State University, University Park, PA 16802

Cultures of *E. coli* CY15070 cells harboring the pWHA43 plasmid (1) encoding for methionyl-prochymosin under the control of the trp promotor showed a constant fraction of cells containing inclusion bodies during exponential growth over a 2000-fold increase in cell numbers. The finding is in accordance with a "pole age" model of cell growth which describes the inheritance of inclusion bodies by daughter cells and explains the morphological changes occurring in the cultures. Cultures inoculated at 140 CFU/mL had only 8% inclusion body containing cells during exponential growth whereas those inoculated at 0.14 CFU/mL had around 40% inclusion body containing cells. The pole age model and direct measurement of the prochymosin content of the cells both indicate that cells inoculated at a low cell density produce more protein than cells inoculated at a high cell density. Contrary to expectations, however, this higher rate of prochymosin production did not result in a lower cell growth rate. Cells inoculated in the range of 0.14 to 140 CFU/mL all grew at the same specific growth rate of 1.67 hr^{-1}. The cultures inoculated at the lower cell densities had a significantly longer lag time (~6 hours) than those inoculated at higher density (~1 hour).

It is now well known that the overproduction of proteins in *E. coli* cells frequently results in the formation of refractile granules called inclusion bodies within the cell envelope. Inclusion bodies are amorphous particles consisting primarily of unfolded or misfolded agglomerates of the protein overexpressed from the plasmid and other cell components involved in protein synthesis (2). Several recent studies have focused on the nature of these bodies (3-5) and the isolation of the protein contained within them from whole cell lysates (6-9).

0097–6156/91/0470–0133$06.00/0
© 1991 American Chemical Society

Very few reports have commented specifically on the kinetics of the changes in the cell morphology which occur during inclusion body formation and the kinetics of cell growth and expression in these systems. Of those that have (10,11) there is some consensus that when inclusion bodies form, cells elongate and the inclusions appear at one or both ends of the cells and eventually along the length of the cell (11-13). The elongation of cells is not universally observed however, and may be a function of the host cell used as well as the protein being expressed (10,14).

In this report we have examined the kinetics of the appearance of inclusion bodies for a particular inclusion body forming expression system. We propose a general model for the formation of inclusion bodies within a culture as a whole.

Materials and Methods

Bacterial Strain, Plasmid and Growth Media. *Escherichia coli* strain CY15070 (tnaA2 trpR2 lacI^q), a derivative of W3110 (15), was used in all experiments as the host strain. LB media was used in all cultures. Plasmid pWHA43 was obtained from McCaman (1) in a related host cell (CY15001). The plasmid was isolated from theses cells using the CsCl/Ethidium Bromide method as described by Birnboim (16), then transformed into CaCl$_2$-treated CY15070 cells according to the method of Dagert and Ehrlich (17) as described below.

The pWHA43 plasmid contains 3.9 kbp and replicates with the pMB1 origin of replication. The plasmid was constructed with an insert of cDNA encoding met-prochymosin under the control of a tandem lac-trp promotor operator arrangement as described in (1). The prochymosin gene was inserted just upstream from the ampicillin resistance gene without a transcription terminator sequence between, but with the RNA polymerase binding site for amp also intact, so that presumably amp is transcribed from both the lac-trp and natural amp promotors.

According to the original study (1) both the lac and trp promotor sequences control expression of the prochymosin gene. However, studies with lac inducer (IPTG) indicated that the lac promoter had no effect on expression, perhaps due to the fact that it is too far upstream of the gene. The trp control region which appears to be driving expression of the gene had been modified in such a way as to eliminate the attenuator sequence. Since the host cells are tryptophan repressor deficient, the cell is believed to express the proteins in an essentially constitutive manner.

Transformation Procedure. Cells were transformed using the CaCl$_2$ procedure of Dagert and Erhlich (17). CY15070 host cells were grown to an optical density (540 nm) of 0.2 in LB broth, chilled on ice then pelleted at 7000 rpm in a benchtop centrifuge. The supernatant was pipetted off and the pellet was resuspended in 0.2 mL cold 100 mM CaCl$_2$ solution. The suspension was held on ice for 20 minutes. The procedure above was repeated and the cells were then held on ice for 20-24 hours. The next day 0.5 micrograms of pWHA43 DNA was added to the cell suspension. The cells were held on ice for an additional 30 minutes then were incubated at 37 C for about 2 minutes. Two mL of LB media (no ampicillin) was added and the cells were further incubated at 37 C for 2 hours. Aliquots of this culture were mixed with 80% sterile glycerol to a final concentration of 10 v/v % and then were stored frozen at -80 C. These were called the "primary transformants". The transformation efficiency was estimated to be about 1 x 10^{-5} cells transformed per host cell.

Plating Procedures and CFU Determination. Viable cell assays were conducted using a Spiral Systems plating device according to the manufacturer's instructions. Cultures were sampled periodically with sterile pipets, appropriately diluted, then plated onto LB plates or LB plates containing 100 micrograms of ampicillin/mL broth. The plates were incubated overnight at 37 C, then individually counted to determine the colony forming units (CFU). The CFU/mL in the original samples calculated from this data by allowing for the dilution of the original sample. In a few cultures the total cell count was also determined using a Petroff-Hausser cell. In several cultures the CFU/mL were obtained on both selective (amp) and non-selective plates. Within 6 hours after inoculation, the same CFU/mL (within experimental error-LB plates often showed less CFU/mL than the corresponding amp plates as expected from random error) were obtained on both ampicillin and non-selective LB plates, indicating both that the plating efficiency of the antibiotic plates was high and that there were few if any plasmid free cells in the culture.

In all cases the plating procedure gave highly reproducible results, and yielded growth curves which were straight lines on semi-log plots (r>0.99).

SDS Electrophoresis and Gel Scanning. Samples were prepared for SDS-polyacrylamide gel electrophoresis by a lysis-boiling procedure. 2 mL of a culture sample was lysed by sonication with a Branson sonicator and microtip at setting #4 40% duty for 4 minutes. 200 microliters of the sonicate was mixed with 100 microliters of SDS loading buffer (0.3 g SDS, 0.01 g bromophenol blue, 3 mL glycerol, 1.5 mL beta-mercaptoethanol, and 1.88 mL tris-HCl(pH 6.8) per 10 mL). The mixture was boiled for 5 minutes in a microfuge tube then 20-50 microliters were loaded into a lane. The samples were separated on a 10% SDS gel with a 2% stacking buffer (18) for 5 hours at 30 milliamps.

The gels were stained with Coomassie Blue then scanned on an LKB 2222-20 Ultrascan XL Densitometer. Although comparison of the lanes from host and prochymosin producing cells made it obvious that prochymosin was located at a position corresponding to 40 kilodaltons, the protein was also located using the western blotting procedure of Burnette (19) on an identical gel. Sheep anti-rennin from East-Acres Biologicals and anti-sheep IgG-alkaline phosphatase (Sigma) were used to locate the prochymosin band using nitro-blue-tetrazolium and bromochloroindoyl phosphate as indicators.

SDS gels were stained with Coomassie Blue and scanned using an LKB 2222-20 ultrascan XL densitometer. The amount of prochymosin was reported relative to the amount of a constitutively produced protein located just above the prochymosin band. The expression of this protein was determined to be proportional to the overall protein concentration in the cell sample and therefore acted as an internal standard for prochymosin content.

Cell Culture Procedures. All the data reported in these studies were from cultures grown in 125 mL shake flasks containing 25 mL of media in a New Brunswick G24 rotary shaker at 37 $\pm$ 0.2 C and 250 rpm. Some experiments were conducted in well aerated 1 liter fermentors (Lh Fermentation) with similar results as the analogous shake flask experiment (data not shown).

Visualization and Counting of Cells and Inclusion Bodies. Total cell counts were obtained by counting the cells in a Petroff-Hausser cell according to the manufacturer's instructions. The cells were visualized in a Nikon Microscope equipped with a phase contrast lens and aperture rings using the oil immersion lens. The fraction of cells containing inclusion bodies was determined by direct inspection of reactor samples under the microscope. An inclusion body containing cell was scored as one containing a visible green object within the cells. For cell densities below 1 x 10^6 CFU/mL it was necessary to centrifuge the cells to concentrate the samples before enough cells could be seen. For each sample counted between 10 and 100 cells were scored for inclusion bodies. In most cases, 100 cells were counted, but practical considerations prevented this from being possible when less than 1 x 10^6 CFU/mL were present.

Results

Influence of Inoculum Density on Culture Lag Time. Frozen primary transformant cultures were thawed, assayed for CFU/mL on ampicillin plates, then appropriately diluted and inoculated into 125 mL shake flasks containing 25 mL of LB media with 100 μg/mL ampicillin. The dilution of the inoculum was varied in different experiments to give a CFU/mL ranging from 0.14 CFU/mL to a maximum of 140 CFU/mL. Periodic samples of each of the cultures were taken and the CFU/mL were determined as a function of time by plating assays. The growth data were regressed and plotted in a typical semilog fashion. The linear region of the exponential growth curve was then extrapolated back to the CFU/mL in each culture immediately after inoculation as determined from a plate assay of the fresh transformant culture. The time corresponding to this extrapolation was designated as the (apparent) lag time, t_{lag}. The constructions used for determination of t_{lag} in four representative experiments are shown in Figure 1.

The value of t_{lag} determined in this way was found to depend strongly on the size of the inoculum for these cells. At low values of the inoculum size, (0.14 CFU/mL) the lag time was around six hours but it decreased to less than

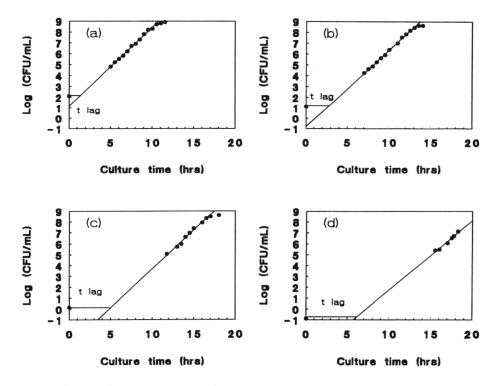

Figure 1. Apparent lag times and experimental growth rates for a series of different inoculum densities. (a) 140 CFU/mL; (b) 14 CFU/mL; (c) 1.4 CFU/mL; (d) 0.14 CFU/mL.

2 hours for high inoculum size (140 CFU/mL). A series of experiments showed that there was a linear correlation between the logarithm of the inoculum size and the apparent lag time for the culture for inocula of between 0.14 and 140 CFU/mL (Figure 2). Host cells grown from frozen cultures did not show any inoculum size effect over the same range of CFU/mL. (It should be noted that the host cells were not subjected to the conditions of the transformation process.)

Influence of Inoculum Density on Culture Growth Rate. The slopes of the growth curve data were used to obtain a value for the maximum growth rate (μ_{max}) of the cells in LB media as summarized in Table I. Only sample cell densities of between 5×10^4 and 5×10^8 CFU/mL were used in the regression in order to avoid lag phase and stationary phase data.

Table I

SUMMARY OF SPECIFIC GROWTH RATE DETERMINATIONS

Inoculation Density (CFU/mL)	μ^1	n^2	t_{lag}(hrs)
140	1.68 ± 0.08	11	1.36
14	1.66 ± 0.05	12	2.70
6	1.66 ± 0.08	9	4.95
1.4	1.64 ± 0.14	8	5.03
0.6	1.75 ± 0.04	9	7.40
0.14	1.47 ± 0.24	6	5.90

[1]95% confidence limits on value of μ.
[2]n is the number of contiguous points used in the regression.

There appeared to be no significant difference in maximum growth rate for cultures inoculated at higher densities than those inoculated at lower densities (Figure 3). Cells inoculated at 0.14 CFU/mL to 140 CFU/mL grew at a rate corresponding to a μ_{max} averaging about 1.67 hr^{-1}. This can be compared to a measured host cell growth rate of 2.2 hr^{-1} and the growth of a low producing mutant cell containing pWHA43 with a putative downshift promotor mutation which was 2.0 hr^{-1} ([14]).

Changes in Cell Morphology When Inclusion Bodies are Expressed. As soon the cell density was high enough to allow for visualization of individual cells under the microscope (10^5 - 10^6 cells/mL), samples of each of the cultures were taken and the fraction of cells containing visible inclusion bodies was determined

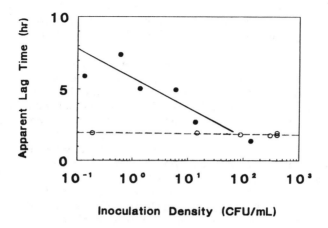

Figure 2.   Apparent lag time as a function of inoculum density for CY15070/pWHA43, (•) and CY15070 (○) cells.

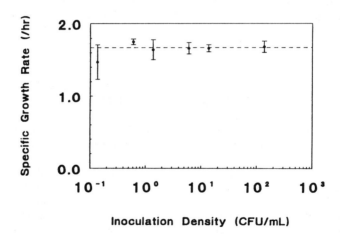

Figure 3.   Specific growth rates of CY15070/pWHA43 as a function of inoculum density.

in the phase contrast microscope as described in the Materials and Methods section.

In each culture the fraction of cells containing visible inclusion reached an approximately constant value during exponential phase growth (Figure 4). As the cells reached late exponential or early stationary phase the fraction then rose sharply so that eventually 90-95% of the cells contained at least one inclusion body.

The general appearance of the cells at various culture times are sketched in the insert box in Figure 4. The inoculated cells were mainly host cells and therefore contained no visible inclusions. By the time the culture reached 1 x 10^5 CFU/mL a fraction of the cells had one visible inclusion which appeared as a distinct small dark spot or greenish-blue object located at one pole of the cell. In low inocula the inclusion bodies appeared to be larger than in high inocula throughout exponential phase. Some of the cells were two to three times the normal length in the lower inocula but most were only slightly elongated in higher inocula. As the cells reached stationary phase (~5 x 10^8 CFU/mL) inclusions began to appear in more than one end of nearly all of the cells. Further into stationary phase, cells which were elongated 2-10 times normal length appeared. About 20-30% of these cells contained multiple inclusion bodies, usually distributed at regular intervals along the cell length.

Influence of Inoculum Density on the Fraction of Cells Containing Inclusion Bodies. The fraction of cells containing inclusion bodies was constant during exponential growth for a single inoculum, but depended strongly on the density

Table II

SUMMARY OF FRACTION OF CELLS WITH VISIBLE INCLUSION BODIES

Inoculation density (CFU/mL)	Number of cells observed	Number of cells with inclusions	Fraction[1,2] with inclusions
140	431	36	0.084 ± 0.025[a]
140	442	43	0.097 ± 0.024[a]
14	305	44	0.144 ± 0.039
6	205	48	0.234 ± 0.029
1.4	180	63	0.350 ± 0.036[b]
0.6	250	90	0.360 ± 0.030[b]
0.14	150	63	0.420 ± 0.040
0.14	551	200	0.363 ± 0.039[b]

[1]95% confidence limits calculated from binomial distribution
[2]Fractions with same letter not statistically different at 95% level.

of the inoculum used (Figure 5 a-d). At low inoculum densities (0.14 CFU/mL) around 40% of the cells contained visible inclusions during exponential growth when the cell density was between 1×10^5 CFU/mL and 2×10^8 CFU/mL. This fraction decreased sharply as the inoculum density was increased until for an inoculum density of 140 CFU/mL only around 8% of the cells showed visible inclusions. When all the measurements of the fraction of cells containing visible inclusions taken during exponential phase were pooled for a given inoculum density the differences between the average fraction between inoculum densities were significant at the 95% level between most samples (Table II). There was a roughly linear correlation between the fraction of cells containing inclusions (during exponential growth) and the logarithm of the inoculum density (Figure 6).

As the cell density increased to 2×10^8 CFU/mL there was a rapid increase in the fraction of cells containing inclusion bodies in all cultures. Regardless of inoculum density, all cultures reached about 90-95% of the cells containing at least one inclusion in stationary phase.

Influence of Inoculum Density on the Prochymosin Content of the Culture. Culture samples were taken at various times during a culture and the prochymosin content was determined using Coomassie Blue stained SDS polyacrylamide gels and quantitative scanning as described in the Materials and Methods section. To minimize measurement error inherent in the staining process and to avoid ambiguities associated with trying to scan the entire gel, the prochymosin peak area was scaled relative to an unknown protein appearing in gels run from host cells. The band was roughly a constant relative intensity as compared to all the cell proteins and therefore was assumed to be proportional to the cell mass.

The densitometer measurements show that the fraction of cells containing visible inclusion bodies correlated to the (relative) amount of prochymosin in the culture as a whole (Figure 7). This correlation appeared to hold both within a single culture as it entered stationary phase and between different cultures with different inoculum sizes . The correlation was fairly linear until about 80% of the cells contained inclusion bodies then the amount of prochymosin sharply increased relative to the other protein band. This corresponded to early stationary phase when there was a sudden increase in prochymosin content and an appearance of multiple inclusion bodies within each cell.

Discussion

Experience with the CY15070/pWHA43 cells had shown that subculturing of the cells was difficult because of the formation of a mutant cell incapable of forming a visible inclusion body (14). Once the cells were plated (LB w/amp), picking a colony and subculturing into fresh antibiotic containing media resulted in cultures containing a much smaller fraction of inclusion body containing cells when the cultures were fully grown. In some cases no inclusion bodies were seen

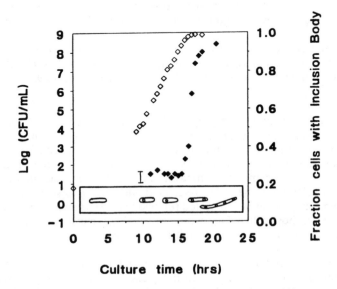

Figure 4.   Growth (◊) and inclusion body expression (♦) for a typical culture. The inoculum density was 6 CFU/mL.

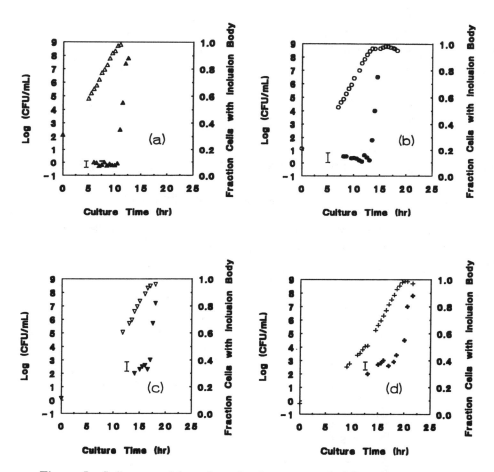

Figure 5.   Influence of inoculum density on growth (open symbols) and inclusion body expression (closed symbols) in CY15070/pWHA43: (a) 140 CFU/mL; (b) 14 CFU/mL; (c) 1.4 CFU/mL; (d) 0.6 CFU/mL.

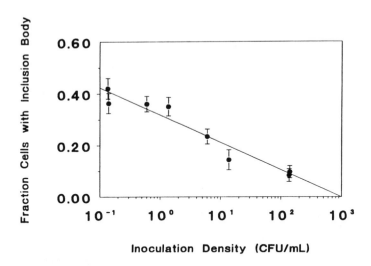

Figure 6. Influence of inoculum density on the fraction of cells containing inclusion bodies during exponential growth (10^5 - 10^8 CFU/mL).

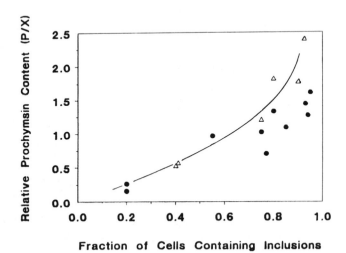

Figure 7. Relationship between the fraction of cells containing inclusions during exponential growth and the relative prochymosin content of the cells. Inoculum density 0.14 CFU/mL (▲); inoculum density 140 CFU/mL (•).

in fully grown cells in these plate-to-liquid subcultures. This behavior was traced to the formation of a mutant cell which retained a plasmid but did not produce visible inclusions. SDS acrylamide gels showed that this mutant produced only about 10% of the prochymosin as did fresh transformants. The mutant also produced about 10% of the amount of beta-lactamase of the fresh transformants. The plasmid content of the mutant was similar to the plasmid content of fresh transformants suggesting that it was a downshift mutant in the trp promotor influencing expression of both the prochymosin and co-transcribed beta-lactamase.

Serial liquid-to-liquid subculture experiments were run to determine the reason for the rapid takeover by non-inclusion body-forming cells. When subcultures were taken from stationary phase cells, the fraction of inclusion body containing cells in the population in stationary phase dropped dramatically to essentially zero within two subcultures. When subcultures were obtained from exponentially growing cells, the fraction of cells showing inclusion bodies in stationary phase dropped much more slowly with each subculture, 84% of the stationary phase cells showed visible inclusions even after four subcultures.

Our conclusion from these experiments was that CY15070/pWHA43 cells which reached stationary phase had very low viability, both in liquid and on agar plates. Because of this, subculturing resulted in selection for a more viable non-inclusion body forming cell producing less prochymosin. Subculturing from plates gave a similar result so that good product yields were not possible in subculture or plate-to-liquid cultures.

Because of these subculturing problems a protocol was adopted which called for using primary transformants as the inocula for batch growth studies. Each culture was started by inoculating a flask or reactor with an aliquot of cells taken directly from the -80 C freezer. It was assumed that the ampicillin in the reactor would appropriately select for plasmid containing cells and cause lysis of the non-transformed hosts still contained in the primary transformant culture media. This approach gave satisfactory results at small scale, ie the cultures grew rapidly and 90-95% of the cells contained inclusion bodies in stationary phase, however, attempts to scale-up by inoculating larger reactor sizes resulted in much longer total culture times than one would calculate from small scale data by simply accounting for the increased dilution of the inoculum. It appeared that the lag time was much longer in the large reactor, probably because the relative size of the inoculum was smaller. We decided to explore this phenomenon more thoroughly at small scale by using different inoculum sizes to initiate the cultures. Anomalous results were obtained in several respects as outlined in the Results section.

The results presented here can be summarized as follows:

1) The apparent lag time of the culture depended strongly on the inoculum size. Small inocula resulted in significantly longer lag times than larger inocula.

2) Once the cell density reached 10^5 CFU/mL or larger, cultures from different starting inoculum sizes grew at a similar specific growth rate, about 75-80% of the host cell growth rate in the same media.

3) The fraction of cells which contained visible inclusion bodies during exponential growth was constant but depended on the inoculum size. Larger inocula yielded cultures with significantly fewer inclusion body containing cells. The amount of prochymosin correlated to the number of cells showing visible inclusions.

4) All the cultures, regardless of inoculum size, resulted in at least 90% of the cells showing at least one inclusion body in stationary phase.

<u>Lag time effects</u>. The relationship between the apparent lag time and the inoculum density observed in these experiments seems to be the result of a biphasic growth pattern in the batch cultures. This is seen more clearly in a composite growth curve for several different inoculum sizes as shown in Figure 8. To make this plot, a constant delay time was added to the culture time values for each data set to maximize the overlap of all the growth curves. This manipulation of the data demonstrates that overall growth appears to involve two periods, a slow growth period when the cell density is less than 1×10^5 CFU/mL and a rapid growth period when the cell density is between 1×10^5 and 1×10^8 CFU/mL. The data can be fit by assuming a simple 2-cell model of this process where a higher number of slow growing cells initially present in the culture are overgrown by faster growing cells. The cell specific growth rate is about 0.8 hr^{-1} until the cell density reaches 10^5 CFU/mL and then increases to a higher specific rate of about 1.7 hr^{-1} between 10^5 and 10^8 CFU/mL.

The log dependence of the apparent lag time on the inoculum density can be completely explained by this growth pattern. Low inocula had longer apparent lag times because they grew for a longer time at a low initial growth rate. Since the lag time was determined by extrapolation, assuming that they always grew at the higher growth rate, this would lead to the pattern observed.

The transition from a slow growing to fast growing cell is apparently related to the cell density since all inocula fit on a common curve. If there were a contaminant of "fast growing" cells in a culture of "slow growing" cells, one would expect that domination of the reactor by the fast growing cells would occur at the same time after inoculation, not at the same cell density. It is possible to explain the result by assuming that the appearance of a fast growing cell requires a rare mutation event, and that the probability of the mutation is small until the cell density reaches a minimum value, but this scenario is not proven by the data. In any event, whatever the cause, there appears to be a culture density dependent transition to a faster growing cell. Visual inspection of the cells about the time of the increase in the cell growth rate indicated that the cells are longer than later in exponential phase. One may therefore speculate that the slow growth rate observed is due to the increased cell size. If the rate of increase in mass of the cell between divisions is constant, then it follows that the specific growth rate, determined by an increase in cell numbers, will be smaller when the cells grow larger before cell division.

We found that the apparent lag time had no relationship to the size of the reactor inoculated but only depended on the inoculum density. (Data not shown.)

Constant Fraction of Cells Containing Inclusion Bodies During Exponential Growth. A constant fraction of inclusion body containing cells is a consequence of exponential growth and the fact that inclusions are inherited by daughter cells as discrete objects. This can be shown by consideration of the cell growth and division process (Figure 9). For the purpose of this analysis a cell can be considered to be divided into three distinct regions, and "old" pole, a "new" pole, and a "nuclear" region ([20]). The "old" pole "new" pole concept comes from the fact that cell wall synthesis takes place in the center of the cell, pushing old wall material to the end. Since the cell also divides in the center, at birth there will be an old and a new pole region as defined by the age of the cell wall material. This process of growth and division continues as long as the cells are growing exponentially, so that there will be a certain fraction of cells with one pole age equal to the number of generations the culture has been growing, a fraction with one pole equal to an age one generation less than the number of generations the culture has been growing and so forth. In synchronous cells the fraction of the cells with a given pole age at birth is given by,

$$P(m,n) \; = \; \frac{1}{2^m} \qquad m < n$$

$$P(m,n) \; = \; \frac{1}{2^{m-1}} \qquad m = n$$

Where $P(m,n)$ is the fraction of cells of with the older of the two poles of age m in a culture growing exponentially for n generations. For example, for an exponentially growing culture that has been dividing for 3 generations one would expect to find 1/2 of the cells with pole age (0,1), 1/4 of the cells with pole age (0,2) and 1/4 of the cells with pole age (0,3). Furthermore, because the culture is growing exponentially, a constant fraction of the cells will be of pole age (0,m) or older regardless of the number of generations a culture has been growing.

Extending the pole age concept to the cytoplasm region (since there is no completely free exchange of material from one side of the nuclear region to another--ie the nuclear material partitions the cell), then in terms of synthesis of protein for the inclusion body material, one end of the cell will be older than the other. This end will have had longer to accumulate an inclusion body than the newly formed end. This suggests that when it takes a certain number of

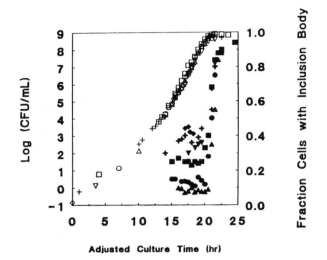

Figure 8. Master growth and inclusion body expression curves for inocula of
0.14 - 140 CFU/mL: 140 CFU/mL (△,▲); 14 CFU/mL (○,•); 6
CFU/mL (□,■); 1.4 CFU/mL (▽,▼); 0.6 CFU/mL (+,+); and 0.14 CFU/mL
(◊,♦).

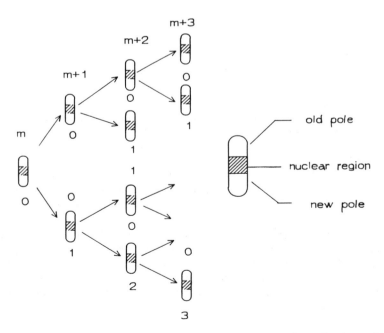

Figure 9. Schematic representation of the pole age hypothesis and the
appearance of visible inclusion bodies in exponentially growing cells.

generations for a visible inclusion body to form, a constant fraction of the total population of cells will contain a visible inclusion at any time. The inclusion bodies in the ends of the very oldest cells will be larger than those in the youngest cells and one would expect to see various sized inclusion bodies within a population of cells.

This model for inclusion body formation explains why only one inclusion body forms per cell. Under the assumption of exponential growth, all of the cells always have one pole between 0 and 1 generation old. If one generation is not enough time to synthesize enough protein to form a visible inclusion then one cell pole must be free of inclusion bodies.

Inoculum Density Versus Fraction of Cells with Inclusion Bodies. Cultures from higher inocula showed a significantly smaller fraction of cells containing visible inclusion bodies than inocula from low inocula. The reason for this is not clear at this time but it not due to the presence of plasmidless cells in the culture. Due to the low transformation efficiency, the inoculum contained 10^5 times as many plasmidless cells as plasmid containing cells. However, the former were rapidly killed in the growth medium. By 6 hours into the culture there were less than 1% plasmid free (antibiotic sensitive) cells in the culture. No plasmidless cells were detected by comparative plating on selective and non-selective plates at the cell densities where the inclusion body fraction was determined.

The pole-age model proposed above explains why the fraction of inclusion body containing cells is constant within a culture during exponential growth but does not explain why lower inocula resulted in a higher fraction of cells with inclusion bodies than high inocula. It does, however, suggest that the relative synthesis rate is higher in low inocula than in high inocula. From our results one may calculate the number of generations required to form a visible inclusion body in a cell. For low inocula, when 40% of the cells show a visible inclusion, about 2.25 generations are required for a visible inclusion to form. For high inocula, where 8% of the cells have visible inclusions, it takes 5 generations for a visible inclusion to form. One may also conclude that the net rate of protein synthesis is higher in the cultures with a higher fraction of inclusion body containing cells. This seems to be verified in the prochymosin measurements shown in Figure 7.

Influence of Inoculum Density on Cell Growth Rate. Surprisingly, despite the difference in the fraction of inclusion body containing cells in the different cultures studied here, there appears to be no difference in the specific cell growth rate during exponential growth for different sized inocula. As argued above, the different inclusion body fractions imply different protein synthesis rates. One might expect that if this was the case then there should also be different specific growth rates for the cells since more cell protein synthesizing resources are directed toward non-essential prochymosin production in the higher producing cells. Put another way, the inoculum size appears to influence the balanced growth *state* of the cells without influencing the balanced growth *rate*.

Currently we can offer no tested explanation for this anomaly. One possibility which has yet to be tested is that both cultures are synthesizing prochymosin at the same rate, but that the degradation system is more active in the higher inocula. Under this hypothesis, low inocula have a higher net rate of protein synthesis but not a higher rate of denovo synthesis of prochymosin. This would explain the difference in inclusion body formation and at the same time explain the similar growth rates observed in high and low inocula cultures but leaves open the question of why higher inocula would have more active protein degradation systems.

Accumulation of Inclusion Bodies in Stationary Phase. When the cells begin to enter stationary phase there is a sudden increase in the fraction of cells with visible inclusions. This is followed by the formation of cells with multiple inclusions. Similar findings have been observed in other studies. Cheng ([21]) observed that visible inclusions were produced only in the stationary phase in *E. coli* cells producing a truncated beta-galactosidase molecule, even though synthesis was occurring in exponential phase cells as well. Schoemaker et al ([22]) made observations which also imply that inclusion body formation occurred primarily in late exponential- early stationary phase in *E.coli* B/r containing the pCT70 plasmid which encodes for expression of prochymosin. Schoner et al ([11]) observed first one, then two, then multiple inclusion bodies in *E.coli* RV308 cells harboring the pCZ101 plasmid encoding for bovine growth hormone.

Formation of more than one inclusion body is consistent with the pole-age model proposed. As the cells enter stationary phase, cell division is halted and (according to the model) there is no new cell wall synthesis. If, however, protein synthesis persists in the cytoplasm, then recombinant protein will accumulate. In effect, the cell poles will be getting older without an influx of young poles and inclusion bodies will appear in the aging ends of the cells already in the culture. Cell elongation and the formation of more than two inclusion bodies in a single cell may be the result of overexpression of recombinant proteins causing interference with the cell division system. When this is the case, multiple nuclear regions may form in the elongated cells, effectively segmenting them into cytoplasmic regions of different ages. Eventually the cytoplasmic regions between nuclear regions accumulate enough protein to demonstrate a visible inclusion body.

Summary

The observations made in this study on inclusion body formation are explained by a pole age model of cell growth. The constant fraction of cells with inclusion bodies during exponential growth is a consequence of segmentation of cell cytoplasmic regions by the nuclear material in the cell, and the fact that inclusion bodies are discrete objects which cannot migrate from one end of the cell to another. The model predicts that the size of inclusion bodies will depend on the number of generations a culture has been expressing the protein and the specific synthesis rate. The model is consistent with the observations that at most one inclusion body will appear in exponentially growing cells (under most conditions) and that multiple inclusions will appear in cells entering the stationary phase.

The pole age model does not explain why low inoculum densities resulted in a higher fraction of the cells with visible inclusion bodies and higher synthesis rates for prochymosin without decreasing the exponential specific growth rate in the cells.

Acknowledgments

A portion of the funding for this work was supplied by the National Science Foundation under grant EET-8613138. We express our gratitude to Dr. Joseph Hall of the Pennsylvania State University for assistance with the electrophoresis analysis.

Literature Cited

1. McCaman, M. T.; Andrews, W. H.; Files, J. G. *J. Biotech.* 1985, *2*, 177-190.
2. Hartley, D. L.; Kane, J. F. *Biochem. Soc. Trans.* 1988, *16*, 101-102.
3. Fish, N. M.; Hoare, M. *Biochem. Soc. Trans.* 1988, *16*, 102-104.
4. Kane, J. F.; Hartley, D. L. *Topics in Biotechnology* 1988, *6*, 95-101.
5. Taylor, G.; Hoare, M.; Gray, D. R.; Marston, F. A. O. *Bio/Tech.* 1986, *4*, 553-557.
6. Marston, F. A. O.; Lowe, P. A.; Doel, M. T.; Shoemaker, J. M.; White, S.; Angal, S. *Bio/Tech.* 1984, *2*, 800-807.
7. Williams, D. C.; Van Frank, R. M.; Muth, W. L.; Burnett, J. P. *Science* 1982, *215*, 687-689.
8. Marston, F. A. O. *Biochem. J.* 1986, 240, 1-12.
9. Masui, Y.; Mizuno, T.; Inouye, M. *Bio/Tech.* 1984, 2, 81-85.
10. Fieschko, J.; Ritch, D. B.; Fenton, D.; Mann, M. *Biotech. Prog.* 1985, *1*, 205-208.
11. Schoner, R. G.; Ellis, L. F.; Schoner, B. F. *Bio/Tech.* 1985, 2, 151-154.
12. Marston, F. A. O.; Angal, S.; Lowe, P. A.; Chan, M.; Hill, C. R. *Biochem. Soc. Trans.* 1988, *16*, 112-115.
13. Yasueda, H.; Nagase, K.; Hosoda, A.; Akiyama, Y.; Yamada, K. *Bio/Tech.* 1990, *8*, 1036-1040.
14. Reddy, J. M.S. Thesis, Penn State University, 1989.
15. Paluh, J. L.; Yanofsky, C. *Nuc. Acids Res.* 1986, *14*, 7851-7860.
16. Birnboim, H. C. *Meth. Enzymol.* 1983, *100*, 243-255.
17. Dagert, M.; Ehrlich, S. D. *Gene* 1979, *6*, 23.
18. Laemmli, U. K. *Nature* 1970, *227*, 680.

19.   Burnette, W. N. *Anal. Biochem.* 1980, *112*, 195-198.
20.   Helmstetter, C. E.; Leonard, A. C. *J. Mol. Biol.* 1987, *197*, 195-204.
21.   Cheng, Y-S. *Biochem. Biophys. Res. Comm.* 1983, *111*, 104-111.
22.   Schoemaker, J. M.; Brasnett, A. H.; Marston, F. A. O. *EMBO J.*, 1985, *4*, 775-780.

RECEIVED March 26, 1991

Chapter 12

Active Creatine Kinase Refolded from Inclusion Bodies in *Escherichia coli*

Improved Recovery by Removal of Contaminating Protease

Patricia C. Babbitt[1], Brian L. West[2], Douglas D. Buechter[1], Lorenzo H. Chen[1], Irwin D. Kuntz[1], and George L. Kenyon[1]

[1]Department of Pharmaceutical Chemistry and [2]Metabolic Research Unit, University of California—San Francisco, San Francisco, CA 94143

We have expressed creatine kinase (CK) from *Torpedo californica* electric organ cDNA in *E. coli*. Although the expression system was designed to produce native, cytoplasmic protein, the initial protein product is insoluble, with no detectable activity. It can be isolated in an enriched form by centrifugation and is stable to proteolysis for several months when stored at 4 °C. Since soluble CK from tissue sources can be denatured and refolded with nearly complete recovery of activity, we adapted this methodology for use with the expressed CK in order to obtain soluble, active protein. Only minute amounts of active protein were recovered using this procedure, however, and further exploration of this phenomenon showed that this was due to proteolysis of the aggregated CK during denaturation and refolding. The CK is apparently resistant to this proteolysis in the aggregated state but is highly sensitive to it during denaturation and refolding. While this proteolytic activity is inhibited by neither phenylmethane sulfonyl fluoride (PMSF) nor EDTA, it can be largely removed either by extraction with a detergent-containing buffer prior to denaturation of the CK or by ion exchange chromatography under denaturing conditions. These methods result in 30-100 fold improved recoveries of soluble protein which, upon further purification, yield pure protein which is highly similar to CK isolated from tissue sources. Experiments with extracts containing this proteolytic activity show that it is largely inactive on native soluble CK, degrading this protein only after denaturation and refolding are initiated. We have also found a proteolytic activity contaminating inclusion bodies formed when a mutant of bovine pancreatic trypsin inhibitor (BPTI) is expressed in *E. coli*. It is similar to that found in the CK system, acting to degrade expressed protein only after the initiation of denaturation and refolding procedures. We therefore suggest that this proteolytic activity may be a general phenomenon complicating the refolding of other inclusion body proteins as well.

The formation of inclusion bodies reflects a general and often intractable problem associated with expression of eukaryotic DNA's in prokaryotes. Even though significant amounts of insoluble protein can be isolated, solubilization by denaturation and refolding often results in low recoveries, particularly when refolding is attempted

0097–6156/91/0470–0153$06.00/0

on aggregates which are contaminated with endogenous proteins (1-4). While possible explanations have focused on sub-optimal folding conditions or concentrations or on the presence of contaminating proteins released by solubilization procedures, there is little direct evidence to support any specific cause for these low recoveries. Nevertheless, there has been considerable progress in the development of methods for increasing the recovery of aggregated proteins (5-8), and there is an increasing number of reports of recovery of active soluble protein from such expression products (3, 4, 9-12). Most of these methods involve the purification of the aggregates to remove contaminating proteins at some point prior to the initiation of refolding.

We have recently reported that removal of contaminating proteins associated with inclusion bodies produced during the expression of both CK and BPTI in E. coli also produces a dramatic increase in the yield of refolded enzyme (13). Our investigations showed that these contaminating proteins exhibit a proteolytic activity which is quiescent in the aggregate, acting to degrade expressed proteins only after solubilization by denaturation/refolding procedures. The "delayed action" aspect of this phenomenon made it difficult to discern initially, and we suggest that this may contribute to its being overlooked in other systems as well. This report includes a description of the expression of CK as an insoluble aggregate in E. coli, initial characterization of the proteolytic activity associated with it, and an evaluation of methods tested for removing this proteolytic activity and for obtaining soluble active protein by refolding.

Materials and Methods

Materials. Electrophoresis reagents were obtained from Bio-Rad, Emeryville, CA. Ampholytes were from Pharmacia, Piscataway, NJ. Guanidine-HCl (ultrapure) was purchased from Schwartz-Mann, Cleveland, OH, and urea (ultrapure) from International Biotechnologies, New Haven, CN. The n-octyl-ß-D-glucopyranoside (octyl glucoside), rabbit muscle CK and other reagents were obtained from Sigma, St. Louis, MO. Ultrafiltration cells, YM30 membranes, Centricon and Centriprep units for protein concentration were obtained from Amicon, Danvers, MA. Plasmid BPTI$_{694}$ was a gift from Mr. John Altman, University of California, San Francisco, CA. Expression plasmid pKT52 was a gift from Dr. Karen Talmadge, California Biotechnology, Palo Alto, CA. Sequencing reagents were from United States Biochemical Corp., Cleveland, OH. Oligonucleotide primers for both mutagenesis and dideoxy sequencing were either synthesized by the Biomolecular Resource Center, University of California, San Francisco, CA or obtained as a gift from California Biotechnology, Inc. Bacterial strains were acquired from Bethesda Research Laboratories, Bethesda, MD (JM101), from Dr. P. Greene, University of California, San Francisco, CA, (D1210), and from Dr. M. Inouye, State University of New York, Stony Brook, NY (JA221). Full-grown *Torpedo californica* was obtained from Pacific Bio-Marine Laboratories, Venice, CA.

Purification of CK Isolated from the Electric Organ of *Torpedo californica*. Cytoplasmic CK was purified from *Torpedo californica* electric organ by modification of the scheme used by Barrantes *et al.* (14) as described by Buechter (15). Following isolation of the cytoplasmic CK from electric organ tissue, the enzyme was purified by EtOH precipitation and affinity chromatography on Blue Sepharose. Further purification was obtained using mono Q fast protein liquid chromatography (FPLC) equilibrated with 50 mM Tris-acetate, pH 8.5. CK was eluted with a NaCl gradient.

Expression and Recovery of Soluble CK from Insoluble Inclusion Bodies. Construction of the Expression Vector, pKTCK3F. The cDNA insert encoding CK from the electric organ of *Torpedo californica* was isolated from pCK52g8 (*16*) and cloned into an expression vector containing a hybrid *trp/lac* promoter (*trc*) (*17*). This vector, pKT52, is a modification of vector pKK 233-2 (*18*) from which the *EcoRI/PvuII* fragment has been deleted to produce a high copy number vector of 2895 bp. The *NcoI* cloning site in pKT52 was changed to *SphI* and an *SphI* site was introduced at the initiating ATG in the CK clone by site-directed mutagenesis (*19*). Creating the matching *SphI* sites allowed linkage of the CK coding sequence to the optimized bacterial promoter in a way that yields the natural CK without any modification of the second translated amino acid (proline). To make a compatible cloning site for the 3' *BgIII* site in the noncoding region of pCK52g8, a *BgIII* linker was ligated into the *HindIII* site of the *SphI*-modified pKT52. The 1239 bp CK insert was isolated from the *SphI-modified* pCK52g8 following digestion with *SphI* and *BgIII* and ligated into the *SphI/BgIII*-modified pKT52. Sequencing of the mutated regions was done in bacteriophage M13 by the dideoxy method (*20*). Sequence verification of the promoter and CK-coding regions of the final construction, named pKTCK3F, was also done by the dideoxy method optimized for double-stranded templates (*21*). Because we found that the plasmid CK expression vector could be prepared using bacterial strain D1210 without the occurrence of mutations which abolish CK expression, constructions were made in this strain (*22*).

Expression and Isolation of CK. In order to obtain high copy number expression, the CK was expressed in strain JA221 (JA221/F'*laci*q) (*23*). The cells were transformed with pKTCK3F and plated onto LB agar containing 40 µg/ml ampicillin and grown at 37 °C overnight. A freshly transformed colony was transferred to a starter culture of 30 ml minimal medium (*24*) containing thiamine-HCl, tryptophan, leucine, and 40 µg/ml ampicillin. This culture was grown overnight with shaking (300 rpm) at 37 °C and transferred to 1 l fresh minimal or LB medium. At an OD_{650} of 0.2-0.8, protein synthesis was induced by the addition of isopropyl-β-D-thiogalactoside (IPTG) to 1 mM final concentration. Cells were grown for four more hours and harvested by centrifugation at 3000 x g. This and subsequent procedures were done at 4 °C except as specified.

The cell pellets were weighed and resuspended in either 50 mM Tris or in 50 mM Tris containing 8% sucrose, 5% Triton X-100, 50 mM EDTA (STET) (*25*) to a volume of 10 ml/g wet weight of cells. These and other Tris-containing buffers used in subsequent procedures were adjusted to pH 8.5. To each 10 ml of suspension, 130 µl of 10 mg/ml lysozyme (freshly prepared) was added and the suspensions incubated overnight. The suspensions were then sonicated 3 times for 3 minutes each with the energy set at the tip maximum and the efficiency level set at 50%. A pulsed signal was used. The pellets were collected by centrifugation at 12,000 x g for 30 minutes and resuspended in 50 mM Tris. The suspensions were sonicated and centrifuged as above, and this process was repeated until no more protein was observed in the supernatants. Samples were resuspended in 50 mM Tris to an estimated CK concentration of 1 mg/ml.

Extraction of Aggregated CK with Octyl Glucoside and other Detergents. The aggregated CK isolated from the Tris lysis buffer was suspended to an estimated CK concentration of 0.5 mg/ml in 50 mM Tris containing 2.5% octyl glucoside and, for some experiments, 5% ß-mercaptoethanol. This suspension was incubated at 37 °C with slow shaking (~100 rpm) for >16 hours and centrifuged at 12,000 x g for 20 minutes. The pellets were washed 2 times with 50 mM Tris and resuspended at an approximate CK concentration of 1 mg/ml. Alternatively, extraction of the pellet was also attempted by a two-hour incubation at room temperature followed by overnight extraction at 4 °C with 1% Triton X-100, 20% glycerol, 2% deoxycholate in 50 mM Tris (*26*).

Denaturation and Refolding of CK Denaturation and refolding methods were modifications of procedures used for refolding soluble CK isolated from tissue sources (27). CK pellets were centrifuged and resuspended in 50 mM Tris at an estimated CK concentration of 2 mg/ml. Solid guanidine-HCl or urea, solid dithiothreitol (DTT), and 50 mM Tris were added so that the final guanidine-HCl or urea, DTT, and CK concentrations would be 6 M or 8 M, 100 mM, and 1 mg/ml, respectively. The samples were left for 20 minutes to 2 hours at room temperature, then diluted 60-fold into 50 mM Tris with rapid swirling. The dilute refolding mixture was left overnight at 4 °C. Prior to analysis, the refolded CK was concentrated ~60-fold either by ultrafiltration or by Centriprep or Centricon centrifugation using either 30,000 or 10,000 MW cutoff membranes.

Purification of Expressed CK on FPLC in 8 M Urea. Expressed CK aggregates isolated in both Tris and STET and containing approximately 1 mg CK each were centrifuged for 12,000 x g for 2 minutes and the supernatant removed. The pellets were resuspended in 200 μl 50 mM Tris, pH 8.5, to which 192 μg solid urea and 8 μl 1 M DTT were added. The samples were thoroughly mixed and volumes were brought to 400 μl with buffer. The samples were left at room temperature for ~ 1 hour, and 150 μl was removed to a new tube for refolding without further purification. The remainder was chromatographed on a Mono Q FPLC column (Pharmacia) equilibrated with buffer A and running at 0.9 ml/min., with the following gradient. Buffer A was 50 mM Tris-acetate containing 8 M urea, pH 8.5 and Buffer B was Buffer A containing 1.0 M NaCl.

Time (min.)	% Buffer B in the Gradient
0	6
10	6
30	8
40	8
40	75
50	75

FPLC Purification of Expressed CK's Following Refolding. Expressed, STET-extracted CK aggregate and tissue-purified CK were refolded as described. Aliquots of these samples were filtered and chromatographed on a Mono Q FPLC column equilibrated with Buffer A at the rate of 1.0 ml/min., with the following gradient. Buffer A was 50 mM Tris-acetate, pH 8.5 and Buffer B was Buffer A containing 1.0 M NaCl.

Time (min.)	% Buffer B in the Gradient
0	0
10	0
15	4
20	4
25	10
35	10
45	75

Incubation of Octyl Glucoside Extracts of the CK Aggregate with Soluble Rabbit Muscle CK. The octyl glucoside extract obtained from the extraction of the CK aggregate was tested for proteolytic activity on soluble rabbit muscle CK. Approximately 13 mg of CK aggregate was extracted with octyl glucoside as described above. This extract (23 ml) was concentrated to 1 ml, and the supernatant was twice exchanged by ultrafiltration with 20 ml of 50 mM Tris to remove most of the detergent. Within 24 hours, a precipitate formed which was removed by centrifugation at 12,000 x g for 8 minutes. The pellet was washed twice with 1 ml of 50 mM Tris and resuspended in 0.5 ml of buffer. Aliquots of the resuspended pellet and the supernatant (ranging from 0 to 133 μl) were incubated with 300 μg aliquots of soluble rabbit muscle CK prepared at 10 μg/μl in 50 mM Tris. The samples were brought to equal volumes with 50 mM Tris and 10% of each mixture was removed and stored at 4 °C for later analysis. To the remainder, buffer, solid urea, and 1 M DTT were added to give final concentrations of 8 M urea and 100 mM DTT, respectively, in 300-400 μl total volumes. The mixtures were left for 2 hours at room temperature and refolded as described above. The effects of PMSF and EDTA as protease inhibitors were tested as follows: PMSF was dissolved in isopropanol and added to the refolding buffer to achieve a final concentration of 0.1 mM and EDTA was added to a final concentration of 0.1 mM.

Expression and Recovery of Soluble BPTI from Insoluble Inclusion Bodies. Expression and Isolation. An $Asn_{43} \rightarrow Leu_{43}$ mutant of BPTI,, $BPTI_{694}$, was expressed in *E. coli* strain MM294 (*28*) under the control of a *trp* promoter. It does not contain a signal sequence but was engineered to express as a fusion protein within the cytoplasm (*29*). Cells were grown at 37 °C, 350 rpm, in M9 medium containing thiamine, 0.5% glucose, 0.5% casamino acids and 15 μg/ml tetracycline to an $O.D._{650}$ of 1.0, then induced for 4.5 hours with 3-ß-indoleacrylic acid (25 μg/ml). Cells were harvested and suspended in either 50 mM Tris or STET. Lysis and isolation of the insoluble pellets were performed as described for CK.

Denaturation and Refolding of BPTI. Pellets isolated either in 50 mM Tris or in STET and containing approximately 1 mg of BPTI were denatured in 100 mM Tris, 1 mM EDTA, 25 mM DTT, 8 M urea and left at room temperature for 30 minutes. Based on Creighton's conditions (*30*), refolding was initiated by 60-fold dilution into 50 mM Tris, 5 mM reduced glutathione, 1 mM oxidized glutathione, 50 mM KCl. The refolding solutions were stored at 4 °C overnight, centrifuged to remove insoluble material, and concentrated on Centricon concentrators using a 3,000 MW cutoff membrane.

Miscellaneous Methods. Sodium dodecyl sulfate-polyacrylamide gel electrophoresis (SDS-PAGE) was performed by standard methods including heat treatment of the samples in Laemmli buffer containing β-mercaptoethanol prior to electrophoresis (*31*). Gels were stained with Coomassie Blue G-250. Isoelectric focusing was performed using native conditions at 4 °C according to the ampholyte manufacturer's directions. Protein determinations were made either by the methods of Lowry (*32*) or of Bradford (*33*). CK activity was determined by the method of Tanzer and Gilvarg (*34*). The presence of full-length CK was evaluated by SDS-PAGE. N-terminal sequencing of CK purified from *Torpedo californica* electric organ tissue was performed by the Biomolecular Resource Center, University of California, San Francisco.

Results

Purification of CK from the electric organ of *Torpedo californica*. Partially purified CK isolated from electric organ tissue was chromatographed on FPLC as described, and the major CK-containing fraction was submitted for N-terminal sequencing. This procedure showed that the first 37 amino acids were the same as those predicted by the cloned KTCK3F cDNA sequence (*16*). This material was used for comparison with expressed CK in subsequent experiments. It had a specific activity of 19.3 U/mg of protein.

Expression and Yield of CK from *E. coli*. Expressed CK was recovered exclusively as an insoluble aggregate under all growth conditions examined (Figure 1, lane 3). No CK activity could be detected in these inclusion bodies and neither full-length CK (evaluated by SDS-PAGE) nor CK activity could be detected in the supernatant fractions. The yield of CK and the proportion of the insoluble aggregate that it represents are sensitive to a number of variables, including the culture medium, the density of cultures at the time of induction, total induction time, and whether the protein is expressed in small or large-scale cultures. Both the yield and the proportion of CK in the inclusion bodies were greater for cells cultured in Minimal rather than LB media following transformation onto LB agar. When cells were grown to greater than 0.2-0.6 O.D.$_{650}$ units prior to induction, expression of CK decreased in both nutrient media. Compared to induction times of less than 6 hours, overnight induction also resulted in large proportional losses of CK expression. Small-scale preparations (50 mL or less) produced more CK per O.D.$_{650}$ unit of cells than did large-scale growth (1 L or more). Growth conditions and induction times were therefore optimized empirically to produce the best yields of expressed protein. The wet weights of harvested cells ranged from 2-6 g/l of culture, and the yield of CK in the insoluble pellets was estimated at 1-3 mg/g wet weight of cells.

Solubilization and Refolding Studies. Initial attempts to recover active CK from the aggregated expression product were directed at solubilizing the protein. In response to reports suggesting that inclusion bodies can be solubilized by adsorption to ion exchange resin and recovered by salt gradient elution, we tried to solubilize aggregated CK with both Fast Q and Fast S Sepharose (Pharmacia) (10). Neither CK activity nor CK protein (as judged by SDS-PAGE) was detectable in any of the supernatant fractions collected over an eluent range of 0-2 M NaCl. Thus, the CK aggregate was not solubilized using this method.

The CK was also invariably insoluble in the initial lysis buffer (50 mM Tris, pH 8.5 or 50 mM Tris, pH 8.5 containing 5 mM EDTA) as well as in two other buffers tested: 50 mM Tris, pH 8.5 containing1% Triton X-100, 20% glycerol, 2% deoxycholate and 50 mM Tris, pH 8.5 containing 2.5% n-octyl glucoside, 5% β-mercaptoethanol. In contrast, addition of either solid guanidine-HCL or urea directly to the pellets dispersed in lysis buffer (to produce final denaturant concentrations of 6M and 8M, respectively) completely solubilized the pellets. On the other hand, addition to centrifuged pellets of either 6M guanidine-HCl or 8M urea solutions gave only incomplete solubilization. The "greasy" appearing residue not solubilized by treatment of the pellet with either denaturant solution was soluble in 2X Laemmli buffer (31), however, and, when run on SDS-PAGE, appeared to be composed of the same material as that in the insoluble pellets prior to denaturation.

These studies suggested to us that the only practical method for recovery of soluble CK was by complete denaturation and refolding. Initial attempts to refold the expressed CK, however, resulted in >95% losses of full-length CK protein and specific activities no better than 2% of that of soluble *Torpedo* electric organ CK purified from tissue and denatured and refolded as a control. Refolding either by dialysis or 60-fold dilution produced a flocculent precipitate. Slightly better recoveries of soluble CK were obtained from refolding by dilution, and this method was used routinely thereafter.

Purification of Insoluble CK Prior to Refolding.   Extraction with Detergent. Based on these results, we suspected that protein contaminants in the pellet were interfering with refolding. We therefore tried extracting the CK pellet with detergents in an attempt to solubilize and remove the contaminants. Extraction of the pellet by the non-ionic detergent, n-octyl glucoside, removed significant amounts of contaminating protein from the pellet (Figure 1, lane 4). When these detergent-extracted pellets were denatured and refolded, a 30-fold improvement in recovery of active CK was observed. In addition, the formation of precipitate observed during refolding of the unextracted pellet was greatly reduced. Based on the success of the octyl glucoside extraction, we tested other methods to purify the aggregated CK and further increase the yield of refolded CK.

Replacement of the original lysis buffer (50 mM Tris) with STET (50 mM Tris containing 8% sucrose, 5% Triton X-100, 50 mM EDTA) (*25*) produced the best results. The purification achieved by this extraction is shown in Figure 1, lane 5. The CK pellet isolated from STET appears to be as pure as that isolated from 50 mM Tris followed by octyl glucoside extraction (Figure 1, lane 4). The STET extraction procedure has two other advantages: it is less expensive to use than octyl glucoside, and it reduces the number of steps in the purification scheme.

In order to quantitate the improvement in recovery due to purification of the aggregates prior to refolding, the three pellets shown in Figure 1, lanes 3-5, were refolded under identical conditions. The recoveries and specific activities were analyzed as shown in Figure 1, lanes 7-9, and in Table I.

**Table I.   Comparison of Refolded Expressed
CK Purified by Various Methods**

Sample Name	U/mg	% U/mg of Control	% Recovery of Protein	Units Recovered
Control CK (Purified from tissue)	20.2	100	59	12.0
Expressed CK:				
Untreated aggregate	0.04	0.2	63	0.003
Octyl glucoside-extracted aggregate	0.90	4.7	52	0.5
STET-extracted aggregate	5.6	29.0	63	3.5
Untreated aggregate, FPLC purified in 8 M urea	14.6	75.6	0.42	0.6
STET-extracted aggregate, FPLC purified in 8 M urea	16.0	82.9	0.36	0.6

CK was denatured and refolded as described. The experiments were done at the same time and denaturing and refolding conditions were adjusted to make them as identical as possible for each sample. Approximately 1 mg of CK was denatured in each case. The amount of starting material denatured as "Control" CK was measured by Lowry analysis of CK purified to homogeneity. The amount of starting material denatured as "Expressed" CK was estimated from densitometry measurements of CK-containing insoluble aggregates fractionated on SDS-PAGE and stained with Coomassie Blue.

For comparison, CK purified from *Torpedo* electric organ tissue was denatured and refolded under the same conditions as the expressed aggregates (Figure 1, Lanes 2, 6 and Table I). Protein refolded from pellets that had been purified by extraction with either octyl glucoside or STET could be recovered in reasonable yield as shown in Table I. Both of these pellets and the protein refolded from them appear to be of similar purity, although STET treatment improves both recovery and specific activity of the refolded CK more than does the octyl glucoside extraction. In contrast, the protein refolded from the unextracted pellet was so extensively degraded that neither enzyme activity nor full-length protein was observed (Figure 1, lane 7), even though the amount of protein recovered in this sample appeared similar to that recovered from the detergent-extracted aggregates.

When the untreated CK aggregates are refolded and analyzed on high percentage SDS-PAGE (17.5%), the disappearance of full-length CK correlates with the appearance of new, lower molecular weight bands, confirming the presence of proteolytic activity (Figure 2, lanes 1,2). In contrast, most of the full-length CK seen prior to denaturation in lane 3 is recovered from refolding the STET-extracted aggregate and there are no obvious degradation products (lane 4), leading to the conclusion that STET extraction removes the proteolytic activity.

Anion Exchange Chromatography in 8 M Urea. As an alternative method of purification, the pellets isolated in either STET or Tris were also subjected to anion exchange FPLC following denaturation in 8 M urea. The recoveries and specific activities of the refolded CK purified by this method are also reported in Table 1. Based on these small-scale refolding experiments, the purification method which produced the best specific activities for the refolded CK was anion exchange FPLC in 8 M urea prior to refolding. This method yielded protein of similar specific activity from both unextracted and extracted pellets. These data suggested that purification of the denatured pellets by anion exchange FPLC is even more efficient in removing the protease contaminants that either the octyl glucoside or STET extraction. There was a major drawback in the use of this method, however--losses of CK approached 90% of the total units injected onto the system. Therefore, we decided to attempt large-scale refolding experiments on the STET-isolated pellets without further purification prior to refolding.

Purification of CK Following Refolding. Expressed CK isolated from STET and native CK isolated from electric organ tissue was denatured, refolded, and further purified by anion exchange FPLC (Figure 3). In order to reduce losses of CK which might have arisen partly from the small-scale of our initial refolding experiments, at least 3 mg of CK was processed in each of these experiments. CK purified from *Torpedo* electric organ eluted as fraction 2 in Figure 3A. The comparable FPLC trace for CK expressed in *E. coli*, extracted with STET, and refolded prior to FPLC can be seen in Figure 3B. The region of the trace representing material with CK activity is essentially identical to that in Figure 3A with the exception of the extra peak in fraction 2b. Fractionation of this sample by SDS-PAGE showed that this peak is a non-CK contaminant, but because of its presence, fractions 2b and 2c could not be recovered as pure CK. Of 3 U of CK activity injected onto the column, 1.08 U, or 35%, was recovered in fraction 2a. Approximately 11% of the activity applied to the column was lost in the impure fractions 2b and 2c. The majority of the other proteins contaminating the refolded CK were eluted in peak 3. No protein could be detected on SDS-PAGE in fractions 1a and 1b in Figure 3B. Fifty-one percent of the activity from the CK isolated from electric organ and injected onto the column was recovered as pure CK (fraction 2, Figure 3A).

Figure 4 shows the purity of the expressed CK recovered in fraction 2b. The total recovery of this material can be normalized to 570 μg/l of culture. The specific activity of the CK refolded from the STET-purified pellet and then purified on FPLC (19.0 U/mg of protein) compares favorably with tissue-isolated CK treated in the same manner (20.3 U/mg of protein). The CK from the two preparations exhibit the same pI values and fractionation pattern on isoelectric focusing gels (data not shown).

Characterization of the Octyl Glucoside-Extracted Proteolytic Activity. The proteolytic activity of the octyl glucoside extract was tested on soluble rabbit muscle CK. Most of the enzyme activity of native rabbit muscle CK was maintained after three days of incubation with varying amounts of this extract (Table II). In contrast, when aliquots of these mixtures were subjected to denaturation and refolding procedures, CK activity was lost roughly in proportion to the amount of extract added. These data also show that the precipitate which formed when the octyl glucoside extract was initially concentrated did not exhibit proteolytic activity. Control experiments containing 2.5% octyl glucoside in the buffer showed no loss of CK activity due to denaturation and refolding in the presence of the detergent.

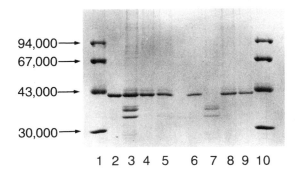

Figure 1. Comparison of untreated and detergent-extracted aggregates and the soluble proteins refolded from them by 10% SDS-PAGE. The amounts loaded in each lane would be equivalent for both insoluble aggregates and soluble proteins refolded from those aggregates if no CK had been lost during the refolding procedure. MW markers (numbered on the left) (Lanes 1 and 10); native CK (Lane 2) and refolded CK protein purified from the electric organ of *Torpedo californica* (Lane 6); insoluble aggregate (Lane 3) and refolded soluble proteins from untreated expressed CK aggregate (Lane 7); insoluble aggregate (Lane 4) and refolded soluble proteins from expressed CK aggregate extracted with octyl glucoside (Lane 8); insoluble aggregate (Lane 5) and refolded soluble proteins from expressed CK aggregate extracted with STET (Lane 9).   (Reproduced with permission from reference 13. Copyright 1990 Bio/Technology.)

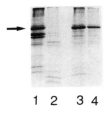

Figure 2. Comparison of refolded CK from untreated and STET-extracted aggregates by 17.5% SDS-PAGE. The amounts loaded in each lane would be equivalent for both insoluble aggregates and soluble proteins refolded from those aggregates if no CK had been lost during the refolding procedure. Insoluble aggregate (Lane 1) and refolded soluble proteins from the untreated expressed CK aggregate (Lane 2); insoluble aggregate (Lane 3) and refolded soluble proteins from expressed CK aggregate extracted with STET (Lane 4). The arrow marks the running position of CK. (Reproduced with permission from reference 13. Copyright 1990 Bio/Technology.)

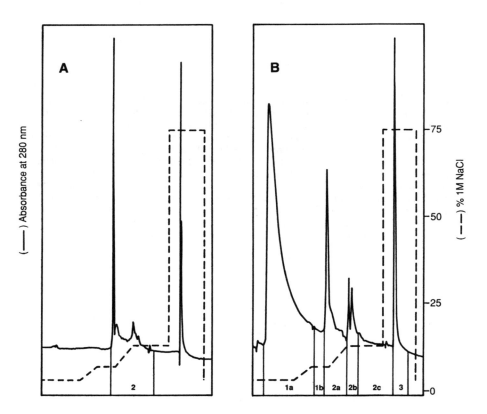

Figure 3. FPLC purification of refolded CK (A) CK isolated from the electric organ of *Torpedo californica* . (B) Expressed CK extracted with STET.

Figure 4. Expressed CK isolated from STET, refolded, and purified on anion exchange FPLC. It has been overloaded on the gel (10% SDS-PAGE) to show the purity of the recovered CK.

**Table II. Proteolytic Effects of the Octyl Glucoside Extract
on Rabbit Muscle CK**

μl extract added	Native % Activity	Refolded % Activity
0	100	82
13	92	65
33	92	20
133	67	<2

"Native % Activity" refers to CK samples which were incubated with extracts for three days, then assayed for loss of enzyme activity. "Refolded % Activity" refers to the same CK samples which were mixed with extracts and immediately denatured and refolded as described, then assayed for loss of activity at the same time as the native samples. CK activity values have been calculated as a percentage of the activity obtained for the native CK which was not incubated with octyl glucoside extract.

Analysis of these samples by SDS-PAGE (Figure 5) show that there is little loss of full-length enzyme for the native CK incubated with varying amounts of either octyl glucoside extract or the precipitate isolated after concentration of this extract. The refolded samples, however, show loss of the CK bands roughly in proportion to the amount of soluble octyl glucoside extract added to the initial mixtures. The other proteins contained in the OG extract which are evident on the gel (lanes 2-4) were not recovered after refolding (lanes 6-8). The samples to which either no octyl glucoside extract was added or only the precipitate which formed after concentration of the extract was added show little loss of CK in the refolded samples. Preliminary results show that neither PMSF nor EDTA are useful in inhibiting this proteolytic activity when added to the refolding buffers at 0.1 mM final concentration. It is unlikely that proteolysis might have occured under denaturing conditions, before these inhibitors were added, since SDS-PAGE analysis of samples left in 8 M urea for three days at 4 °C shows no apparent loss of full-length CK (data not shown).

Discovery of a Similar Proteolytic Activity Associated with the Insoluble Pellet Formed from Expression of a Mutant of BPTI in *E. coli*. Our findings in the CK expression system raised the question whether a similar proteolytic activity could be posing a problem in other systems as well. It was decided, therefore, to test another inclusion body system for similar activity. It has been shown by Altman, *et al.* that several mutants of BPTI can be refolded as active trypsin inhibitor following cleavage of the fusion and affinity purification of denatured material (29). We used their system to produce BPTI inclusion bodies and tested them for the proteolytic activity described herein. We found that a proteolytic activity similar to that associated with CK expression contaminates inclusion bodies produced by expression of one of these BPTI mutants (Asn$_{43}$ → Leu$_{43}$). Similar to our results with CK, we also found that detergent extraction of the insoluble pellet prior to refolding leads to a significant increase in the recovery of full-length protein. The material refolded from the untreated inclusion bodies shows both a loss of refolded BPTI and the appearance of new, lower molecular weight bands (Figure 6, lanes 1, 3). As is also similar to the CK expression system, STET extraction of the BPTI pellet leads to almost full recovery of full-length protein and no apparent degradation products (Figure 6, lanes 2, 4).

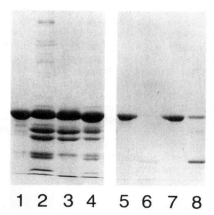

1 2 3 4 5 6 7 8

Figure 5. Comparison of native and refolded rabbit muscle CK incubated with octyl glucoside extracts. Lanes 1-4 (10% SDS-PAGE): 1.2 μl each of native rabbit muscle CK incubated for 3 days without (Lane 1) or with 133 μl (Lane 2) or 33.3 μl (Lane 4) octyl glucoside extract supernatant; 133 μl octyl glucoside extract pellet (material which precipitated out of the octyl glucoside extract when it was washed with 50 mM Tris to remove detergent and concentrated ~ 20-fold) (Lane 3). Lanes 5-8: 2 μl native rabbit muscle CK refolded without (Lane 5) or with 133 μl (Lane 6) or 33.3 μl (Lane 8) octyl glucoside extract supernatant; 133 μl octyl glucoside extract pellet refolded from the precipitated material described for Lane 3 (Lane 7).

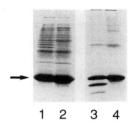

1 2 3 4

Figure 6. Comparison of refolded BPTI from untreated and STET-extracted aggregates on 17.5% SDS-PAGE. Untreated aggregate (Lane 1) and STET-extracted aggregate (Lane 2); refolded soluble proteins from untreated aggregate (Lane 3) and from STET-extracted aggregate (Lane 4). The amounts of BPTI loaded onto lanes 1 and 2 are approximately equal. The amounts of BPTI loaded onto lanes 3 and 4 were calculated to be approximately equal if all of the BPTI had been recovered from refolding as full-length protein. Because the insoluble pellets and the refolded proteins were electrophoresed at different times, the running distances for each set is slightly different. They have been aligned such that the running position of BPTI in both gels is marked by the arrow at the left. (Reproduced with permission from reference 13. Copyright 1990 Bio/Technology.)

Discussion

Initial attempts to isolate expressed CK produced inclusion bodies which, upon denaturation and refolding, resulted in nearly complete loss of full-length CK. The possibility that other cellular proteins contaminating these aggregates might be interfering with the refolding process led us to try to purify the pellets prior to refolding. The results of these experiments were encouraging, leading to as much as 100-fold improvement in the recovery of refolded CK. What surprised us, however, was the magnitude of the improvement. Why did partial removal of contaminants produce such a dramatic increase in both specific activity and product recovery? We were also puzzled that while protein concentration assays showed recovery of equal amounts of protein from refolding both extracted and unextracted pellets, SDS-PAGE analysis of the same samples showed enormous losses of full-length CK from the unextracted pellet and only moderate losses from the extracted pellets (Figure 1, lanes 7-9 and Table I). These observations led us to consider the possibility that the low recoveries of CK from our initial refolding experiments were due to digestion by proteases associated with the contaminating proteins in the pellet (35).

The loss of full-length CK accompanied by the appearance of lower molecular weight bands on the gel confirms that proteolysis during refolding is occurring. The results of our experiments with the octyl glucoside extract supports the conclusion that detergent extraction removes one or more of these proteases from the CK aggregate. These proteases are apparently inactive in both the aggregate and with native, soluble CK. It is only when attempts are made to recover soluble protein by denaturation and refolding that they act to degrade the CK. While initial efforts to inhibit this proteolysis with PMSF or EDTA were not promising, we are currently testing these inhibitors over a range of concentrations as well as other compounds in the hope of finding an effective combination of inhibitors. If successful, we expect that recoveries of refolded protein could be considerably higher, thereby transforming this expression system into one that would be useful on the preparative scale.

Although STET appears to be the best extraction buffer for purification of the CK pellet, the specific activity of CK refolded from this material was still considerably lower than that for the tissue purified control, indicating the need for further purification. The use of anion exchange FPLC in 8 M urea provided this additional level of purification, and samples refolded from this material have specific activities that are comparable to those of CK refolded from tissue-purified controls. We conclude that purification in the denatured state by FPLC apparently removes the contaminating protease(s) and other proteins from the aggregate even more efficiently than either STET or octyl glucoside extraction.

Purification of expressed proteins in the denatured state has been used successfully by other workers (5). In our case, however, there was a serious drawback to this method: total protein recoveries were low compared to CK refolded directly from STET-isolated pellets. We have not yet discovered the reasons for these losses, but because of them we attempted larger-scale refolding experiments directly on STET-purified pellets without the intervening chromatography step on the denatured material. The order of the purification procedures was changed so that the ion exchange chromatography step was performed after, instead of before, refolding. This method produced approximately 6-fold greater recovery of high specific activity CK than when chromatography of the denatured protein was performed before refolding.

Conclusion. We have expressed eukaryotic CK in *E. coli* and have demonstrated that active protein can be recovered from the resulting inclusion bodies. Based on the parameters tested, this protein appears to be very similar to CK purified from electric organ tissue. The recovery scheme is simple, involving cell lysis and overnight extraction with a detergent-containing buffer followed by sonication and isolation of the CK pellet by centrifugation. The pellet is then denatured, refolded and purified to homogeneity by anion exchange chromatography. While the methods we have used to recover active CK are similar to those generally useful in the recovery of active proteins from inclusion bodies, the discovery of the proteolytic activity contaminating the aggregated CK provides new insight into the problem. Finding a similar proteolytic activity associated with the BPTI inclusion bodies suggests that it may complicate expression of other recombinant proteins as well. If so, it may help to explain why purification either prior to or during denaturation has led to improved recovery of such proteins Our laboratory is continuing work directed toward full characterization of the proteolytic activity which, at least in the expression of CK, appears to be a primary component of the aggregation problem.

Acknowledgments

We would like to thank Dr. Charles Craik for help with the mutagenesis of plasmids CK52g8 and KT52 and Dr. Phyllis Kosen for advice on protein denaturation and refolding. We would also like to thank Mr. John Altman for the gift of his BPTI mutant and both Mr. Altman and Ms. Amy Fujishige for contribution of their insights in helpful discussions of the aggregation problem.

This work was supported by U.S. Public Health Service grant AR 17323 (to G.L.K.) and National Institutes of Health grant GM 19267 (to I.D.K.)

Literature Cited

(1)   Cabilly, S.; Riggs, A. D.; Pande, H.; Shively, J. E.; Holmes, W. E.; Rey, M.; Perry, L. J.; Wetzel, R.; Heyneker, H. L. *Proc. Natl. Acad. Sci. USA* **1984**, *81*, 3273-3277.
(2)   Harris, T. J. R.; Patel, T.; Marston, F. A. O.; Little, S.; Emtage, J. S. *Mol. Biol. Med.* **1986**, *3*, 279-292.
(3)   Sharma, S. K.; Evans, D. B.; Tomich, C.-S. C.; Cornette, J. C.; Ulrich, R. G. *Biotech. Appld. Biochem.* **1987**, *9*, 181-193.
(4)   Weir, M. P.; Sparks, J. *Biochem. J.* **1987**, *245*, 85-91.
(5)   Marston, F. A. O. *Biochem. J.* **1986**, *240*, 1-12.
(6)   Marston, F. A. O. in *The Purification of Eukaryotic Polypeptides Expressed in Escherichia Coli*; Glover, D. M.; IRL Press, Oxford, United Kingdom, 1987; pp 59-88.
(7)   Mitraki, A.; King, J. *Bio/Tech.* **1989**, *7*, 690-697.
(8)   Schein, C. H. *Bio/Tech* **1990**, *8*, 308-317.
(9)   Lim, W. K.; Smith-Somerville, H. E.; Hardman, J. K. *Appl. Environ. Microbiol.* **1989**, *55*, 1106-1111.
(10)   Hoess, A.; Arthur, A. K.; Wanner, G.; Fanning, E. *Bio/Tech.* **1988**, *6*, 1214-1217.
(11)   Hanemaaijer, R.; Westphal, A. H.; Berg, A.; Van Dongen, W.; Kok, A. d.; Veeger, C. *Eur. J. Biochem.* **1989**, *181*, 47-53.
(12)   Lin, Z.-l.; Wong, R. N. S.; Tang, J. *J. Biol. Chem.* **1989**, *264*, 4482-4489.
(13)   Babbitt, P. C.; West, B. L.; Buechter, D. D.; Kuntz, I. D.; Kenyon, G. L. *Bio/Tech.* **1990**, *8*, 945-949.
(14)   Barrantes, F. J.; Braceras, A.; Caldironi, H. A.; Mieskes, G.; Moser, H.; Toren, C. E., Jr.; Roque, M. E.; Wallimann, T.; Zechel, A. *J. Biol. Chem.* **1985**, *260*, 3024-3034.
(15)   Buechter, D. D. *Affinity Labeling of Creatine Kinase*; PhD Thesis; University of California: San Francisco, CA, 1990.
(16)   West, B. L.; Babbitt, P. C.; Mendez, B.; Baxter, J. D. *Proc. Natl. Acad. Sci. USA* **1984**, *81*, 7007-7011.
(17)   Amann, E.; Brosius, J.; Ptashne, M. *Gene* **1983**, *25*, 167-178.

(18)   Amann, E.; Brosius, J. *Gene* **1985**, *40,* 183-190.
(19)   Zoller, M. J.; Smith, M. *DNA* **1984**, *3,* 479-.
(20)   Sanger, F.; Nicklen, S.; Coulson, A. R. *Proc. Natl. Acad. Sci. USA* **1977**, *74,* 5463-67.
(21)   Chen, E. J.; Seaburg, P. H. *DNA* **1985**, *4,* 165-170.
(22)   Sadler, J. R.; Tecklenburg, M.; Betz, J. L. *Gene* **1980**, *8,* 279-300.
(23)   Nakamura, K.; Masui, Y.; Inouye, M. *J. Mol. Appl. Gen.* **1982**, *1,* 289-299.
(24)   Maniatis, T.; Fritsch, E. F.; Sambrook, J. *Molecular Cloning: A Laboratory Manual*; Cold Spring Harbor Laboratory: Cold Spring Harbor, New York, 1982; pp 545 .
(25)   Holmes, D. S.; Quigley, M. *Anal. Biochem.* **1981**, *114,* 193-197.
(26)   Babbitt, P. C. *Sequence Determination, Expression, and Site-directed Mutagenesis of Creatine Kinase*; PhD Thesis; University of California: San Francisco, CA, 1988.
(27)   Grossman, S. H.; Pyle, J.; Steiner, R. J. *Biochem.* **1981**, *20,* 6122-6128.
(28)   Meselson, M.; Yuan, R. *Nature* **1968**, *217,* 1110-1114.
(29)   Altman, J.; Henner, D.; Nilsson, B.; Anderson, S.; Kuntz, I. D. *Prot. Engineer.* **1991**, *In press,*
(30)   Creighton, T. E.; Goldenberg, D. P. *J. Mol. Biol.* **1984**, *179,* 497-526.
(31)   Laemmli, U. K. *Nature* **1970**, *277,* 680-685.
(32)   Lowry, O. H.; Rosebrough, N. J.; Farr, A. L.; Randall, R. J. *J. Biol. Chem.* **1951**, *193,* 265-275.
(33)   Bradford, M. M. *Anal. Chem.* **1976**, *72,* 248.
(34)   Tanzer, M. L.; Gilvarg, C. J. *J. Biol. Chem.* **1959**, *234,* 3201-3204.
(35)   Babbitt, P. C.; West, B. L.; Kuntz, I. D.; Kenyon, G. L. *J. Cell Biol.* **1988**, *107,* 184a.

RECEIVED March 8, 1991

Chapter 13

Equilibrium Association of a Molten Globule Intermediate in the Refolding of Bovine Carbonic Anhydrase

Jeffrey L. Cleland and Daniel I. C. Wang

Department of Chemical Engineering, Biotechnology Process Engineering Center, Massachusetts Institute of Technology, Cambridge, MA 02139

Equilibrium refolding of bovine carbonic anhydrase B (CAB) was performed to confirm the observed association of the first intermediate during refolding [Cleland, J. L. and Wang, D. I. C., *Biochemistry*, in press]. The molten globule first intermediate in the CAB refolding pathway was observed to associate at final GuHCl concentrations ranging from 1.8 to 2.2 M and at final protein concentrations greater than 10 μM. Dimer and trimer formation was observed by both quasi-elastic light scattering (QLS) and size exclusion chromatography (HPLC). The dimerization was found to be a slow equilibrium process with an association rate constant of 5.16×10^{-3} min^{-1} and an equilibrium constant of 1.3 μM^{-1}. In contrast, the association for the trimer formation was a rapid equilibrium process. The association rate constant for trimer formation was 0.133 min^{-1} with an equilibrium constant of 0.42 μM^{-1}. Therefore, the equilibrium association of the molten globule first intermediate in the CAB refolding pathway proceeds through the slow dimerization reaction followed by rapid equilibrium association to form the trimer. The association of a molten globule intermediate in both kinetic and equilibrium refolding studies has been observed for other proteins and is probably the common species which causes aggregation during refolding.

The problem of *in vitro* protein folding has become a major barrier to the successful use of bacterial systems for protein production. Bacterial hosts often produce inactive protein in the form of inclusion bodies. The refolding of this inactive protein results in the recovery of the native molecules as well as misfolded and aggregated proteins. Aggregate formation reduces the yield of

0097–6156/91/0470–0169$06.00/0

the native protein and increases difficulties during purification. To refold proteins efficiently in the absence of aggregation, it is usually necessary to perform refolding at low protein concentrations (μg/ml) or at high denaturant concentrations (1 M guanidine hydrochloride or 4 M urea). Alternatively, if the folding pathway is well characterized, the associating intermediate species can be avoided through multiple dilution steps or the addition of an agent which prevents aggregation or enhances refolding (1-2).

Previous work on aggregation during the refolding of bovine carbonic anhydrase B (CAB) and bovine growth hormone (bGH) have shown that an intermediate in the refolding pathway will associate to form multimeric species and can eventually result in protein precipitation (1-2). In particular, bGH associates during equilibrium unfolding at protein concentrations greater than 1.0 mg/ml (3). The associating bGH intermediate has a compact molten globule structure with exposed hydrophobic surfaces (4-5). The formation of a molten globule intermediate has been reported for the refolding of several other proteins (6). Therefore, the association of a molten globule intermediate during refolding is likely to be common for many other proteins.

The protein chosen for this study, CAB, also forms a molten globule intermediate during refolding (7). Equilibrium studies on the unfolding of CAB have shown that the first intermediate has a molten globule structure with exposed hydrophobic surfaces which are not observed in the other intermediate (8). Refolding and aggregation experiments have indicated that this intermediate is responsible for the formation of multimers as shown in Figure 1 (1). To confirm the association of the first intermediate, the equilibrium association of this structure has been extensively studied in this paper. The results from this research yield additional insights into the aggregation previously observed during CAB refolding (1).

Experimental Procedures

Materials. Bovine carbonic anhydrase B (CAB), bovine serum albumin (BSA), guanidine hydrochloride (GuHCl), tris sulfate, ethylenediaminetetraacetic acid (EDTA), and p-nitrophenol acetate (pNPA) were purchased as molecular biology grade reagents from Sigma Chemical Co. (St.Louis, MO). The purity of CAB (pI = 5.9) was checked by high performance liquid chromatography (HPLC) and gel electrophoresis using silver staining. HPLC grade acetonitrile was obtained from J.T. Baker (Phillipsburg, NJ). Decahydranapthalene (Decalin) was a racemic mixture obtained from Aldrich (Milwaukee, WI). All buffers and protein solutions were prefiltered with 0.22 μm disc filters (Millipore Corp., Bedford, MA) or 0.22 μm syringe filters (Gelman Sciences, Ann Arbor, MI). All solutions were prepared using distilled water and further purified with a MilliQ water purification system (Millipore Corp., Bedford, MA).

Protein concentration. The protein concentration of denatured CAB in 5 M GuHCl was determined using a colorimetric dye assay (Bio-Rad Laboratories, Richmond, CA) with bovine serum albumin (BSA) denatured in 5 M GuHCl as the standard.   Assay of protein concentations in 2 M GuHCl were performed by addition of 8 M GuHCl to a final concentration of 5 M GuHCl and analyzed with the BSA standards in 5 M GuHCl.   The extinction coefficient at 280 ηm for CAB in 2 M GuHCl remained constant with protein concentration as observed by absorbance measurements (data not shown).

Quasi-elastic light scattering (QLS) measurements. QLS measurements were performed using a Brookhaven light scattering system (Brookhaven Instruments, Holtsville, NY).   The system consisted of a BI200SM goniometer with a photomultiplier positioned at 90° to the incident laser, 2 W argon ion at 488 ηm (Lexel, Fremont, CA).   The goniometer assembly was temperature controlled using an external water bath.   In addition, the sample was placed in the goniometer with index matching fluid, decalin, surrounding the sample. The photon data was collected using a BI2030 autocorrelator with 128 channels.   A personal computer was used for system control and data storage. The computer was also used to calculate the particle size distributions based on the method of constrained regularization, CONTIN, as described previously (*9*).

Each sample for QLS measurement was equilibrated at the desired final conditions for three to eight hours and maintained in closed containers at 20°C to avoid dust and temperature changes.   After equilibration, each sample was analyzed at least two to three times to ascertain the repeatability of the particle size distribution.   The particle size distributions for each solution showed a high degree of reproducibility.   Multimer distributions were calculated from the particle size distributions as detailed previously utilizing the hydrodynamic radius for 0.10 mg/ml CAB (3.3 μM) in 2 M GuHCl as the persistence length (*1*).

High performance liquid chromatography (HPLC). All HPLC analyses were performed with a model HP1090 analytical HPLC equipped with a diode array detector (Hewlett Packard, Mountain, View, CA).   A Protein PAK 3000 SW column (Waters, Bedford, MA) was used for each size exclusion experiment. The column was equilibrated with ten column volumes (100 ml) of elution buffer (2.0 M GuHCl, 50 mM Tris sulfate, 5 mM EDTA, pH 7.5) prior to operation.   A sample volume of 25 μl was applied to the column and eluted at a flow rate of 1.0 ml/min to facilitate rapid separation.   For equilibrium experiments, each sample was equilibrated for three to eight hours before column separation.

Elution times for native CAB and BSA in the standard dilution buffer (50 mM Tris sulfate, 5 mM EDTA, pH 7.5) were used to calculated the size of the resultant associated species.   In addition, CAB and BSA at low protein concentrations (< 0.10 mg/ml) in 2 M GuHCl were also utilized to determine the effect of denaturant on elution time. The results obtained were comparable to those determined previously (*10*). The elution times were 8.4, 5.6, and 5.1 minutes for monomer, dimer, and trimer, respectively. Since the extinction coefficient at 280 ηm for CAB in 2 M GuHCl was not observed to change with protein concentration, the extinction coefficients of the monomer, dimer, and trimer species were assumed to be equal. With this assumption, the peak areas from absorbance measurements at 280 ηm were utilized to calculate the concentration of each species.

Esterase Activity. Esterase activity was determined as described previously (*1*). The standard for calculating activity was the ester hydrolysis rate constant for the native protein at the same concentration in the dilution buffer (50 mM Tris sulfate, 5 mM EDTA, pH 7.5).

Results

Association Dependence on Denaturant and Protein Concentration. The formation of the first intermediate in CAB equilibrium unfolding experiments has been observed to have a dependence on the final GuHCl concentration (*8*). At GuHCl concentrations ranging from 1.8 to 2.2 M, the first intermediate was highly populated during equilibrium unfolding as measured by binding of a hydrophobic probe, 8-anilino-1-naphthalene sulfonate (ANS) (*8*). Since the first intermediate has been observed to associate during refolding, association of CAB during equilibrium refolding should have the same dependence on the final GuHCl concentration as the observed ANS binding (*1*). Denatured CAB in 5 M GuHCl was rapidly diluted to 33 μM protein and different GuHCl concentrations ranging from 1.0 to 2.5 M. Each sample was allowed to equilibrate for three to eight hours and then analyzed by quasi-elastic light scattering (QLS). The resultant monomer, dimer, and trimer concentrations are shown in Figure 2. The greatest extent of association occurs between 1.8 and 2.2 M GuHCl. These results correlate well with the observed population of the first intermediate determined from ANS binding studies (*8*). Therefore, the first intermediate will associate when present at a concentration of 33 μM during equilibrium refolding. To determine the effect of concentration on this association phenomenon, denatured CAB in 5 M GuHCl was diluted to different protein concentrations ranging from 6.7 to 100 μM and at 2.0 M GuHCl. A final GuHCl concentration of 2.0 M was chosen to maximize the association of the first intermediate as shown in Figure 2. After equilibration, each sample was analyzed by QLS and the distribution of monomer, dimer, and trimer are depicted in Figure 3. The intermediate did not associate at low final protein concentrations ($\leq$ 10 μM). At higher final protein concentrations, CAB associated to form both dimer and trimer species. The multimer distribution shifts to the trimer as the protein concentration is increased to 100 μM. These results were confirmed by HPLC size exclusion chromatography. Each equilibrated sample was analyzed by HPLC as described in experimental procedures and the distribution of monomer, dimer, and trimer are shown in Figure 4. The dependence of association on final

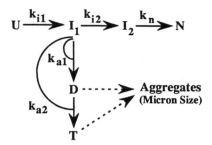

Figure 1: Refolding and aggregation model for bovine carbonic anhydrase B (CAB). When denatured CAB in 5 M GuHCl is rapidly diluted to refolding conditions, the unfolded protein, U, forms the first intermediate, I_1. This intermediate continues to refold to the second intermediate, I_2, which proceeds to fold to the native protein, N. If the unfolded protein is diluted to aggregating conditions, the first intermediate will associate to form multimers. The dimer, D, and trimer, T, species can further associate, resulting in the formation of large aggregates. (Reproduced from ref. 1. Copyright 1990 American Chemical Society.)

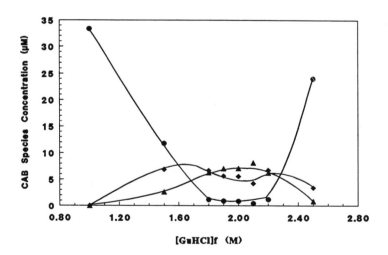

Figure 2: Multimer formation during equilibrium refolding of CAB as observed by quasi-elastic light scattering (QLS). Denatured CAB in 5 M GuHCl was rapidly diluted to a range of final GuHCl concentrations ($[GuHCl]_f$) from 1.0 to 2.5 M and a final CAB concentration of 33 μM and allowed to equilibrate for three to eight hours. The resulting distribution of monomer (●), dimer (◆), and trimer (▲) is shown for each final GuHCl concentration.

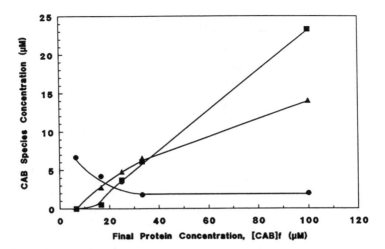

Figure 3: Equilibrium association of CAB as a function of final protein concentration ($[CAB]_f$). Unfolded CAB in 5 M GuHCl was rapidly diluted to a final GuHCl concentration of 2.0 M and a range of final protein concentrations. Each solution was allowed to equilibrate for three to eight hours prior to QLS analysis. The equilibrium monomer (●), dimer (▲), and trimer (■) concentration is shown for each final protein concentration.

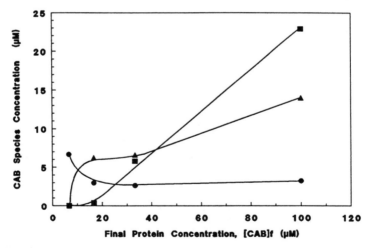

Figure 4: Equilibrium association of CAB as a function of final protein concentration ($[CAB]_f$) as measured by size exclusion HPLC. Equilibrium refolding was performed by rapid dilution of unfolded CAB in 5 M GuHCl to 2.0 M GuHCl and a range of final protein concentrations. After equilibration for three to eight hours, monomer (●), dimer (▲), and trimer (■) concentrations were determined for each final protein concentration as described in Experimental Procedures.

protein concentration as measured by HPLC is similar to the results obtained by QLS. Therefore, the HPLC results do not contain artifacts due to sample dilution or protein-column interactions. The HPLC results were very repeatable with consistent elution times and peak areas. Since the HPLC technique is simpler and yields results similar to QLS, additional experiments were performed using size exclusion HPLC.

From the equilibrium results, a model for the equilibrium association of the first intermediate was postulated for refolding at 2.0 M GuHCl as follows:

$$2 I_1 \rightleftharpoons D \tag{1}$$

$$K_D = [D]/[I_1]^2 \tag{2}$$

$$D + I_1 \rightleftharpoons T \tag{3}$$

$$K_T = [T]/[D][I_1] \tag{4}$$

The first intermediate can associate to form the dimer with an equilibrium constant, K_D, for final protein concentrations greater than 10 μM. The dimer equilibrium constant, K_D, was 1.8 ± 0.2 μM^{-1} as calculated from the QLS data and 1.3 ± 0.1 μM^{-1} using the HPLC data. The largest multimer, trimer, was formed from the association of a dimer and a monomer with an equilibrium constant, K_T. For this association reaction, the equilibrium constant, K_T, was 0.53 ± 0.12 μM^{-1} and 0.42 ± 0.11 μM^{-1} for QLS and HPLC analyses, respectively. The equilibrium constants for dimer and trimer formation are comparable using these two analytical techniques.

Association and Dissociation Rates. With the knowledge of these equilibrium constants, the association rates were determined through additional HPLC experiments. To measure the association rate for dimer formation, denatured CAB in 5 M GuHCl was diluted to association conditions (33 μM protein and 2.0 M GuHCl) and immediately analyzed by size exclusion HPLC. Samples from the refolding solution were analyzed at different times over a period of fifty minutes and converted to concentrations of monomer and dimer as shown in Figure 5. The trimeric specie was not observed over this time period and the relative change in peak areas was significantly different for each measurement. Therefore, the rate constant for dimer formation, k_d, was calculated using the initial rate data for association (Figure 5). The association rate constant for dimerization, k_d, was 5.16 x 10^{-3} min^{-1} with a half time for association of 134.3 minutes. The forward rate constant, k_d, is directly related to the equilibrium constant.

$$K_D = k_d / k_d' \tag{5}$$

Using this relationship, the dissociation rate constant, k_d', was calculated as 3.91×10^{-3} min^{-1} with a resultant half time of 177.3 minutes. In order to measure the rate constants for trimer formation, dissociation experiments were performed by rapid dilution of an equilibrated protein solution at association conditions (33 μM protein and 2.0 M GuHCl) to monomer conditions (3.3 μM protein, 2.0 M GuHCl). The diluted protein solution was analyzed by HPLC size exclusion to follow the dissociation over time (Figure 6). The distribution shifts to the monomer and dimer after ten minutes and the dimer slowly dissociates to form the monomer. The rate constant for trimer dissociation, k_t', was calculated as 0.316 min^{-1} ($t_{1/2}$ = 2.19 minutes) from the rate of decrease in trimer concentration. Again, the relationship between the rate constants and the equilibrium constant was utilized.

$$K_T = k_t / k_t' \tag{6}$$

Based on this equilibrium model, the association rate constant for trimer formation, k_t, was 0.133 min^{-1} with a half time of 5.22 minutes. Since the trimer equilibrium is rapid relative to the dimer equilibrium, the association is difficult to detect at low protein concentrations (< 33.3 μM) using the HPLC technique since this procedure requires a total elution time of 10 minutes and results in a sample dilution of 2.5.

Assessment of Equilibrium Condition. To confirm the achievement of an equilibrium state, the equilibrated solutions were diluted to either the refolding or aggregation conditions which have been observed previously (1). Each equilibrated solution was diluted to a final GuHCl concentration of 1.0 M to allow the protein to refold. After incubation for one hour, the protein completely recovered its biological activity and did not associate to form multimers. These results are comparable to refolding by rapid dilution from 5 M GuHCl to 1 M GuHCl (1). Therefore, the equilibrium associated species can dissociate to allow complete refolding of the protein. On the other hand, dilution of the associated CAB solutions to aggregation conditions (≥ 16.7 μM CAB and 0.50 M GuHCl) resulted in precipitation and recovery of approximately 20% of the biological activity. The precipitation and low recovery of activity were comparable to that observed by direct dilution from 5 M GuHCl to aggregation conditions (1). Therefore, the observed equilibrium association is a true reversible phenomenon.

Discussion

Equilibrium refolding results have shown that the first intermediate in the refolding pathway of CAB will associate to form multimers at high protein concentrations (≥ 10 μM). This intermediate has a molten globule structure with exposed hydrophobic regions (7-8). The molten globule intermediate structure is common in the refolding of other proteins (6). It is therefore likely that the molten globule structure is a general associating intermediate during protein refolding. The association observed for CAB and bGH re-

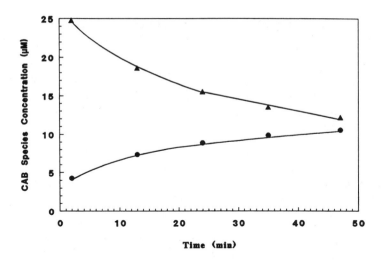

Figure 5: Association of the first intermediate during refolding as measured by size exclusion HPLC. Denatured CAB in 5 M GuHCl was rapidly diluted to association conditions (33 μM CAB, 2.0 M GuHCl) and analyzed at different times for a period of fifty minutes. The monomer (●) and dimer (▲) concentrations are shown as a function of time after dilution to the final conditions.

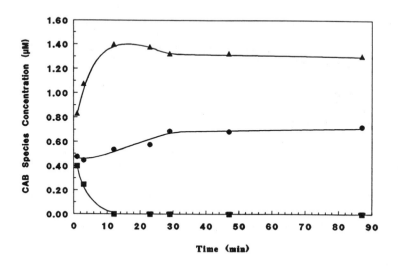

Figure 6: Dissociation of equilibrium associated protein solution. An equilibrium refolding solution at association conditions (33 μM CAB and 2.0 M GuHCl) was diluted to monomer conditions (3.3 μM and 2.0 M GuHCl) and analyzed by size exclusion HPLC. The monomer (●), dimer (▲), and trimer (■) concentrations are shown as a function of time after dilution to monomer conditions.

folding support this hypothesis since both proteins form molten globule intermediates which associate during refolding (*1-2*). In addition, the propensity of this intermediate to form dimers at equilibrium is similar for both proteins (*2*). The formation of a trimer species in CAB equilibrium refolding is strongly dependent on the final protein concentration since the equilibrium is rapid relative to the dimerization reaction. Rapid equilibrium for larger associated species could have also been present in the association of bGH as well as other proteins (*3*).

All protein systems which form this hydrophobic molten globule intermediate are candidates for protein aggregation during refolding at high protein concentrations. Therefore, it is necessary to prevent the association of this intermediate to achieve greater refolding yields. One possible method discussed previously was a two step dilution procedure. First, CAB was refolded at 1.0 M GuHCl to allow the first intermediate to fold in the absence of association. After the second intermediate has been completely formed, subsequent dilution to lower GuHCl concentrations ($\leq$ 1.0 M) which usually result in aggregation allows the formation of native protein in the absence of aggregation (*1*). This approach avoids the pathway of aggregation shown in Figure 1 which occurs for low GuHCl concentrations ($\leq$ 1.0 M) in CAB refolding (*1*). Other methods to avoid aggregation or enhance protein refolding could be achieved by the addition of stabilizing substances such as peptides which have been used to reduce association during the refolding of bGH (*2*). Alternatively, the competing reaction of association could be reduced by enhancing the rate of formation of a nonassociating intermediate on the refolding pathway. The resulting enhanced refolding rate must be greater than the rate of association to achieve a significant decrease in aggregation. Studies are currently underway to develop methods which will either reduce aggregation or enhance refolding.

Acknowledgments

This work was supported by National Science Foundation Grant #CDR-88-03014.

Literature Cited

1. Cleland, J. L.; Wang, D.I.C. *Biochemistry* **1990**, *29*, pp. 11072-11078.

2. Brems, D. N.; Plaisted, S. M.; Kauffman, E. W.; Havel, H. A. *Biochemistry* **1986**, *25*, pp. 6539-6543.

3. Havel, H. A.; Kauffman, E. W.; Plaisted, S. M.; Brems, D. N. *Biochemistry* **1986**, *25*, pp. 6533-6538.

4. Brems, D. N.; Plaisted, S. M.; Dougherty, J.J.; Holzman, T. F. *J. Biol. Chem.* **1987**, *262*, pp. 2590-2596.

5. Brems, D. N.; Havel, H. A. *Proteins: Struct., Funct., Gen.* **1989**, *5*, pp. 93-95.

6. Ptitsyn, O. B.; Pain, R. H.; Semisotnov, G. V.; Zerovnik, E.; Razgulyaev, O. I. *FEBS Lett.* **1990**, *262*, pp. 20-24.

7. Doligkh, D. A.; Kolomiets, A. P.; Bolotina, I. A.; Ptitsyn, O. B. *FEBS Lett.* **1984**, *165*, pp. 88-92.

8. Rodionova, N. A.; Semisotnov, G. V.; Kutyshenko, V. P.; Uverskii, V. N.; Bolotina, I. A.; Bychkova, V. E.; Ptitsyn, O.B. *Mol. Biol.* **1989**, *23*, pp. 683-692.

9. Provencher, S. W. In *Photon Correlation Techniques*; Schulz-Dubois, E. O., Ed.; Springer-Verlag: New York, NY, 1983; pp. 322-328.

10. Corbett, R. J. T.; Roche, R.S. *Biochemistry* **1984**, 23, pp. 1888-1894.

RECEIVED February 6, 1991

Chapter 14

An Engineering Approach to Achieving High-Protein Refolding Yields

Steven Vicik[1] and Eliana De Bernardez-Clark

Department of Chemical Engineering, Tufts University,
Medford, MA 02155

One of the main problems in otherwise efficient recombinant microbial processes is the low refolding yield of biologically active protein from inclusion bodies. A process engineering approach is described to optimize refolding yield using diafiltration as the method of solubilizing agent removal. The strategy employed to optimize refolding yields was to postulate a refolding mechanism, identify and characterize the rate limiting steps in the refolding process, mathematically describe the refolding kinetics, determine the influence of various environmental parameters in the refolding process, and optimize the developed system of equations in a scheme employing diafiltration as the method of denaturant removal. Kinetic equations for competing refolding and inactivation/aggregation reactions formed the basis for mathematically optimizing refolding yield. The refolding of human carbonic anhydrase B (hCAB) from 8M urea was used as a model system to evaluate this approach. A simplified hCAB refolding mechanism is presented together with a technique to characterize the intermediates during hCAB refolding. The processing conditions that maximize refolding yields were mathematically determined and tested using a laboratory scale diafiltration unit.

Advances in genetic engineering have played a major role in the development of new biotechnological processes. Genetic modifications of microorganisms to overproduce eucaryotic proteins frequently result in the accumulation of the protein in insoluble inactive aggregates called inclusion bodies (1). These aggregates can be recovered in the pellet fraction after cell lysis (2). Active protein is commonly obtained by solubilizing the inactive aggregates in a strong denaturant, such as guanidine hydrochloride or urea, followed by refolding the protein to its native state (3). Protein refolding is accomplished by removing the denaturant used to solubilize the inclusion bodies. Pure protein preparations are produced by including purification steps throughout this process (3). In the overall process, the limiting step is frequently the refolding of the denatured protein due to its low yield (4).

[1]Current address: Genetics Institute, 1 Burtt Road, Andover, MA 01810

To date, most studies of protein refolding have been biochemical in nature addressing the questions of why does inactive denatured protein refold to an active form and by what techniques can biologically active proteins be obtained from a denatured inactive species. However, methods for optimizing these refolding techniques have yet to be reported.

The key variables affecting protein stability in the refolding process are temperature, pH, ionic strength, denaturant concentration and protein concentration (5). In proteins stabilized by disulfide bonds in the native state, additional variables include dissolved oxygen, trace metal ion and reducing agent concentrations (6).

A major side reaction in the *in vitro* reconstitution of denatured proteins is the irreversible formation of protein aggregates (7). Proteins aggregate due to intermolecular non-covalent contact formed at the expense of proper non-covalent interactions such as hydrophobic, electrostatic, and hydrogen bonds (8).

Although aggregation is the predominant means by which proteins become inactivated during refolding, several other inactivation pathways have also been observed. Mozhaev and Martinek (8) have described in detail some of these additional methods of protein inactivation. Proteins are inactivated by thiol-disulfide exchange, alteration of the primary structure by chemical modification of amino acid residue functional groups, and cleavage of disulfide bonds. In addition, they may be inactivated by the dissociation of a prosthetic group or the dissociation of an oligomeric protein into subunits.

Aggregation has been observed to occur through a kinetic process with a reaction order greater than two (9). Mozhaev and Martinek (8) suggest that most proteins tend to non-specifically aggregate at protein concentrations on the order of 0.01 to 0.1 mg/ml. Although non-specific aggregation is common, specific aggregation has also been reported (10).

Incubation in solutions of low denaturant concentration can sometimes cause inactivation. Formation of an incorrectly folded form of human carbonic anhydrase in 1.7M GuHCl which subsequently aggregates has been documented (11). Other proteins have also been observed to improperly refold at relatively low concentrations of denaturant. These proteins include ß-galactosidase, tryptophanase, and elastase (6).

It should be noted that low concentrations of denaturant can also favor protein reactivation. For example, incubation in low concentrations of denaturant was observed to promote proper refolding of chymotrypsinogen. The refolding of this protein from 6M guanidine hydrochloride is optimized by diluting to ~1.2M GuHCl in the presence of reduced and oxidized glutathione (12).

Builder and Ogez (13) have patented a similar concept for recovering biologically active protein. They suggested replacing a strong denaturing reagent with a weaker one followed by the reduction in concentration of the weak denaturant. It is thought that this approach preserves the solubility of the refractile protein while subsequently providing a medium which does not interfere with biological activity. This approach is compatible with techniques to properly refold disulfide bonds such as air reoxidation, the addition of reduced and oxidized sulfhydryl reagents, or the addition of sulfhydryl-disulfide combinations. This approach has been used successfully to refold a fusion protein which includes a foot and mouth viral coat protein (13).

It is apparent that while incubation in solutions of low guanidine hydrochloride concentration may sometimes cause inactivation (carbonic anhydrase), it may also promote proper refolding (chymotrypsinogen). In addition, replacement of a strong denaturant with a weaker one is an approach which can be used to successfully refold polypeptide chains. These data suggest that the rate of removal of the denaturant agent can influence yield of active protein. In this respect, time is an important refolding parameter.

This study applies what is known in the biochemical literature about protein folding to develop a process engineering approach to optimizing protein refolding yields. The premise is that refolding yields can be significantly increased over those presently achieved by imposing on the protein an optimal time variant physico-chemical environment in which such parameters as pH, temperature, ionic strength and concentrations of protein and denaturant agents are coordinately controlled in a pre-established manner.

The strategy employed to optimize protein refolding yields involves postulating a refolding mechanism, measuring kinetic and/or equilibrium constants of competing renaturation and inactivation/aggregation reactions, determining the influence of various environmental parameters on the kinetic constants, and optimizing the developed system of equations in a system for denaturant removal with large scale potential. Protein refolding reactions are potentially complex often being multistep in nature. Therefore, applying an engineering approach to optimizing refolding yields includes the identification and characterization of the rate limiting steps in the renaturation process or application of the steady-state approximation to intermediates in the refolding pathway.

Model sytem

The refolding of human carbonic anhydrase B from 8M urea using diafiltration as the method of denaturing agent removal was used as model system to illustrate the engineering approach to optimizing protein refolding yields. The enzyme carbonic anhydrase was selected because of the wealth of biochemical literature available on its renaturation to the native state and its relatively simple refolding pathway. Based on previous studies reported in the literature (14-17) the following renaturation pathway for human carbonic anhydrase B can be postulated:

$$
\begin{array}{ccccccc}
& 1 & & 3 & & 5 & \\
N & \rightleftharpoons & X & \rightleftharpoons & Y & \rightleftharpoons & D \\
2 & & \downarrow 7 & 4 & & 6 & \\
& & I & & & & \\
& & \downarrow 8 & & & & \\
& & A & & & &
\end{array}
\tag{1}
$$

where N is native protein, X is a stable intermediate, Y is a transiently expressed intermediate, D is denatured protein, I is an irreversibly inactivated species and A is aggregated protein.

The kinetic equations can be simplified by applying the pseudo steady-state approximation for Y. The resulting renaturation pathway is:

$$
\begin{array}{ccccc}
& 1 & & 3' & \\
N & \rightleftharpoons & X & \rightleftharpoons & D \\
2 & & \downarrow 7 & 4' & \\
& & I & & \\
& & \downarrow 8 & & \\
& & A & &
\end{array}
\tag{2}
$$

with the following kinetic equations:

$$
\frac{d[N]}{dt} = -k_1[N] + k_2[X]
\tag{3}
$$

$$\frac{d[X]}{dt} = k_1[N] + k'_4[D] - (k_2 + k'_3 + k_7)[X] \tag{4}$$

$$\frac{d[D]}{dt} = k'_3[x] - k'_4[D] \tag{5}$$

$$\frac{d[I]}{dt} = k_7[X] - k_8[I]^n \tag{6}$$

$$\frac{d[A]}{dt} = k_8[I]^n \tag{7}$$

where

$$k'_3 = \frac{k_3\,k_5}{k_4 + k_5} \quad \text{and} \quad k'_4 = \frac{k_4\,k_6}{k_4 + k_5} \tag{8}$$

where the order of the aggregation reaction, n, was determined to be $2.51 \pm .28$ (*18*).

The influence of pH, temperature, and denaturant and protein concentrations on hCAB renaturation yields and aggregation rates, and more specifically, on individual kinetic constants in the refolding pathway was determined. Equations describing the effect of a change in an environmental parameter such as pH, temperature, ionic strength, or organic solute concentration on the refolding rate constants have been proposed in the literature (*19-21*). Based on these functions, the effect of pH, temperature and urea concentration on the kinetic constants for the refolding of hCAB was determined. The results are summarized in Table 1 (*18*).

Refolding yield optimimization. Equations (3) through (7) describe the kinetics of hCAB refolding under a given set of environmental conditions. However, when diafiltration is employed to remove the denaturant, the denaturant concentration is a function of time. For a constant volume, constant filtration rate diafiltration system, the equation describing the rate of denaturant agent removal is:

$$\frac{dC}{dt} = -F/V\,C \tag{9}$$

where C is denaturant concentration. Equation (9) can be integrated between any two times to yield

$$C_2 = C_1 e^{-[F_v\,(t_2 - t_1)]} \quad ; \quad F_v = F/V \tag{10}$$

Equation (10) combined with the equations of Table 1 result in a set of kinetic constants for equations (3) through (7) that are a function of pH, temperature, rate of urea removal (as a function of the diafiltration rate, F_v), and refolding time. By carefully selecting values for each of these variables, the relative rate of the renaturation reactions with respect to the inactivation reactions in pathway (2) can be increased resulting in a higher refolding yield. Selection of values for the process variables is accomplished by solving the following optimization problem:

$$\text{Minimize}\left\{ 1 - \frac{[N]}{[P]} \right\} \tag{11}$$

where [N] is a function of pH, temperature T, diafiltration rate, F_v, and refolding time θ; and [P] is the initial protein concentration (a constant).

Table 1: Effect of pH, temperature and urea activity on kinetic constants

$k_i = A_i e^{-E_i/RT} f_i(a_H)(1 + K'_i a_U)^{\Delta n_i}$					
	A_i	E_i	$f_i(a_H)$	K'_i	Δn_i
k_1	$5.14\ 10^3$	9800	$\dfrac{(a_H + 10^{-7.466})(a_H + 10^{-7.745})}{(a_H + 10^{-5.087})(a_H + 10^{-10.96})}$	0.0146	177
k_2	$8.78\ 10^1$	2190	$\dfrac{(a_H + 10^{-3.183})(a_H + 10^{-4.079})}{(a_H + 10^{-4.595})(a_H + 10^{-6.143})}$	0.403	-3.06
k'_3	$1.15\ 10^{-13}$	-6180	$\dfrac{(a_H + 10^{-7.748})(a_H + 10^{-7.774})}{(a_H + 10^{-6.196})(a_H + 10^{-8.922})}$	51.3	4.16
k'_4	$1.25\ 10^{10}$	12980	$\dfrac{(a_H + 10^{-5.334})(a_H + 10^{-13.47})}{(a_H + 10^{-5.087})(a_H + 10^{-10.96})}$	0.090	-18.4
k_7	$8.46\ 10^{13}$	13880	$\dfrac{(a_H + 10^{-2.582})(a_H + 10^{-9.180})(a_H + 10^{-9.174})}{(a_H + 10^{-8.457})(a_H + 10^{-13.25})(a_H + 10^{-1.863})}$	7720	-1.03
k_8	$1.05\ 10^{10}$	11990	$\dfrac{(a_H + 10^{-6.142})(a_H + 10^{-13.78})}{(a_H + 10^{-7.066})(a_H + 10^{-7.050})}$	0.412	-19.7

T, temperature in K
a_H, activity of H^+ ions
a_U, activity of urea

$$\text{Subject to:} \quad \begin{array}{l} 1 \le pH \le 14 \\ 23°C \le T \le 50°C \\ F_v > 0 \\ \theta > 0 \end{array} \tag{12}$$

[N] is obtained by solving the system of differential equations (3) through (7), with kinetic constants described by the equations in Table 1, and urea activity given by

$$a_U = \gamma(U) [U] ; \quad [U] = [U_0] e^{- F_v t} \tag{13}$$

where [U] is urea concentration, [U_0] is initial urea concentration, and $\gamma(U)$ is urea activity coefficient given by (22):

$$\gamma(U) = 1.000 - 9.142 \ 10^{-2} [U] + 1.409 \ 10^{-2} [U]^2 -$$

$$1.524 \ 10^{-3} [U]^3 + 6.905 \ 10^{-5} [U]^4 \tag{14}$$

Experimental

Materials. Electrophoretically purified human carbonic anhydrase B (hCAB), also referred to as human carbonic anhydrase I, was purchased from the Sigma Chemical Co. (lot #104F93201). Electrophoresis grade urea was purchased from Fisher Scientific. Human carbonic anhydrase B stock solutions were stored at 4°C and used within four weeks, while urea solutions were prepared fresh prior to each experiment

hCAB activity. hCAB activity was assayed by measuring the rate of hydrolysis of *p*-nitrophenyl acetate (*23,24*).

Urea gradient gel electrophoresis. Urea gradient gel electrophoresis was performed based on the method of Creighton (*25*) using a Bio-Rad Mini Protean II gel apparatus. The gradient was prepared by mixing a 15% acrylamide solution containing no urea with an 11% acylamide solution in 8M urea using a gradient mixer. The inverse acrylamide gradient with respect to urea concentration is an empirical correlation determined by Creighton (*25*) to correct for the tendency of urea to slow the migration of all sample components. Gels were poured at 90° with respect to their running position to achieve the desired horizontal urea gradient. Gels were used immediately following polymerization to minimize diffusion of the urea gradient. Human carbonic anhydrase B samples applied to gradient gels were either 0 or 8 M in urea concentration.

hCAB folding kinetic experiments. hCAB samples, at a concentration of 1-3 grams/liter, were denatured in 8M urea and 0.01 M Tris-Cl (pH=7.5) for two to three hours at room temperature (*26*). At the start of an experiment, denatured samples were diluted between 1/10 and 1/20 in a 1 cm path length cuvette placed in a Hewlett Packard 8452A spectrophotometer and the changes in the absorbance spectrum of the sample were recorded every 30 seconds to two minutes. The renaturation medium was 0.09 M Tris-HCl at the appropriate pH, 33 μM $ZnCl_2$, and the appropriate concentration of urea. The diluted hCAB samples had a final volume of 1.4 ml. At various time points over the course of an experiment, 100 μl samples were taken and immediately assayed for concentration of active protein.

In this study, urea gradient gel electrophoresis and denaturing gel electrophoresis were employed to determine conditions under which extinction coefficients for the various intermediate species could be determined. Urea gradient gel electrophoresis (25) is a technique in which a protein sample runs north to south under the influence of voltage difference in an acrylamide gel which has a west to east urea gradient (0M - 8M urea in this study). Applying a native human carbonic anhydrase B sample (0M urea) to such a gel yielded the results presented in Figure 1. It is evident that at low urea concentrations, the protein is entirely in the native conformation. At high urea concentrations, human carbonic anhydrase B is completely denatured and in a random coil configuration (26,35). However, at intermediate urea concentrations, there is at least one intermediate conformation present.

The results obtained from urea gradient gel electrophoresis experiments compare favorably with what has been reported in the literature for the unfolding/folding of carbonic anhydrase. The existence of an intermediate species at 1.2-2M GuHCl or 5M urea has been documented (11,14,16,26,36,37).

The refolding pathway (2) postulates the existence of two intermediates, a conformation which can be potentially reactivated, X, and and irreversibly inactived species, I. There are several reports in the literature that suggest that an irreversibly inactived intermediate is formed during extended incubation at intermediate guanidine hydrochloride or urea concentrations such as 1.7M GuHCl or 5M urea (11,14,26,36,37). In this study, it was observed that human carbonic anhydrase B samples retained ~20% activity after a 90-120 minute incubation in 5M urea (Figure 2) while ~85% reactivation could be obtained after subsequent dilution into 1M urea after one hour. Further incubation in 5M urea resulted in a progressive loss in ability to reactivate samples. The results from these experiments suggest that after a 90-120 minute incubation in 5M urea, hCAB is predominantly in the X conformation. However, after extended incubation (two days to three weeks) at these intermediate denaturant concentrations, the samples are predominantly in the I conformation as determined by the inability to reactivate these samples after dilution to 1M urea. The conformational change from X to I through incubation in 5M urea is illustrated by 5M urea denaturing gel electrophoresis (Figure 3). It is evident that hCAB samples incubated for 90-120 minutes in 5M urea migrate into the gel. However, samples that have incubated from 2 days to three weeks do not effectively migrate into a 5M urea gel.

Based on these findings, extinction coefficients for the various conformational species were measured under the following conditions: 0M urea for native hCAB, 8M urea for the denatured conformer, incubation in 5M urea for 90-120 minutes to estimate the extinction coefficient for X, and incubation in 5M urea for an extended period of time (2-3 weeks was used in this study) to determine the extinction coefficient for conformer I. Initial extinction coefficients evaluations at 238 nm and 292 nm for the various human carbonic anhydrase B conformers were determined in 0.1M Tris-HCl (pH=7.45) at 23 °C. The results are presented in Table 2. It is evident that while there is a substantial (~25-35%) difference in extinction coefficients between the native and denatured states, the intermediate conformations X and I could not be separated spectroscopically, and therefore, UV absorbance cannot be used to determine the concentration of species X and I separately. Instead, the overall concentration of intermediate species, Z where $[Z]=[X]+[I]$, was monitored over the course of a refolding experiment.

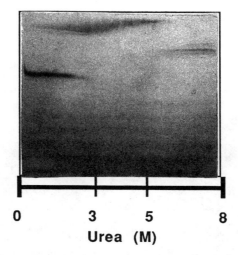

Figure 1. Urea gradient gel electrophoresis, 0 M urea sample, pH = 7.5,
T = 18 - 28°C

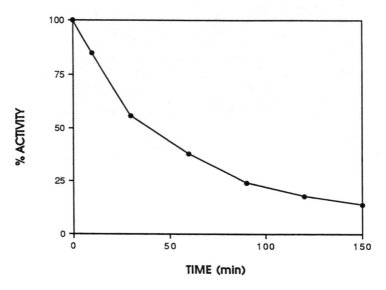

Figure 2. Loss of hCAB activity in 5M urea

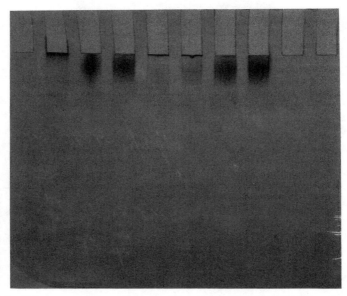

Figure 3. 5M urea gel electrophoresis. N: native hCAB, 5: hCAB incubated in 5 M urea for 90 - 120 minutes, 5*: hCAB incubated in 5 M urea for more than 2 days, 8: hCAB incubated in 8 M urea for 120 -180 minutes. T = 23°C, pH = 7.45

Effect of light scattering on UV absorption spectra. The presence of aggregates in a protein solution cause an increase in optical density in the ultraviolet absorbance region of the spectrum. In order to measure the true absorbance spectra of such solutions, it is necessary to remove the light scattering contribution to the measured optical density in this region. This is accomplished by applying Debye's formulation for scattering and fitting a curve of the form:

$$\Delta D = k\lambda^{-n} \tag{15}$$

between 350 and 800 nanometers (27) and extrapolating this curve into the UV absorbance region. In equation (15) ΔD is optical density difference, λ is wavelength, and k and n are constants. While this technique for separating light scattering effects from true absorbance effects in the UV region is relatively simple, it compares favorably with more complicated techniques such as the one recently described by Garcia-Rubio (28) based on the Mie theory of scattering.

Turbidity measurements. Turbidity was measured with a Hewlett Packard 8452A spectrophotometer at 350 nm. Aggregate concentration was calculated from turbidity measurements using a correlation coefficient of 0.132 g/l absorbance @ 350 nm (18).

Diafiltration experiments. Diafiltration was accomplished using an Amicon spiral cartridge sytem with S1Y3 low protein binding membrane. Protein solution was recirculated through the cross flow device using a Masterflex 7527-34 peristaltic pump. The refolding reactor was a 400 ml polypropylene vessel. Temperature was controlled in the reactor by recirculating water through the outer jacket from a GCA Inc. model 162 water bath. Slight mixing was maintained in the refolding reactor by placing it on a Sybron magnetic stirrer. Fresh buffer was added by maintaining a vacuum in the diafiltration refolding system. UV absorbance spectrum during refolding was monitored continuously by circulating the protein solution through a Hewlett Packard 8452A spectrophotometer using a flow through cell.

Monitoring conformational changes during refolding. In order to determine the concentration of the various conformational species the following techniques were used: UV absorbance at 238 and 292 nm; turbidity, which measures the concentration of aggregated protein [A]; and an activity assay, which measures the concentration of biologically active native protein [N]. Ko *et al.* (29) observed that the recovery of biological activity for bovine carbonic anhydrase B appeared to require complete refolding, therefore, it was assumed that the activity assay would give a direct measurement of the concentration of native conformation, [N].

UV absorbance at a given wavelength results from the contribution of the absorbances, at that wavelength, of all the conformational species present in the sample. In order to use UV absorbance to monitor changes in conformational species concentration, it is therefore necessary to know the extinction coefficients (ε) of the various species. For proteins which reform disulfide bonds during the refolding process, the reaction can be essentially frozen by the addition of iodoacetic acid which stops the conversion of free thiols and essentially arrests the refolding process (30). Intermediates can subsequently be identified electrophoretically (31). However, human carbonic anhydrase B contains only one cysteine residue (32), and disulfide bonds are not formed during the refolding process. Therefore, the technique described above for estimating extinction coeffcents could not be employed in this study. Alternatively, size exclusion, ion-exchange, and hydrophobic interaction HPLC techniques have been reported to be effective in separating protein conformers formed under denaturing conditions (33,34).

Table 2:   Extinction coefficients at 23°C and pH = 7.45

λ (nm)	N	D	X	I
238	2.63±.06	1.76±.03	2.37±.04	2.26±.04
292	1.08±.02	0.82±.02	1.00±.02	0.98±.02

Using the measurements of Z with time during renaturation, the initial and final concentrations of X and I, and the postulated refolding pathway and mathematical model, the concentration of X and I could be estimated during refolding. It was assumed that at the beginning of a dilution refolding experiment, human carbonic anhydrase B was entirely in the denatured state due to previous incubation in 8M urea. Evaluation of concentrations X and I at the end of a dilution refolding experiment was based on activity and electrophoretic data as previously described (Figure 3).

In summary, the concentration of the different conformational species were determined as follows: i) [N] from the activity assay, ii) [A] from turbidity measurements, iii) [Z] and [D] were determined using UV absorbance at 238 and 292 nm with the following equations:

$$\varepsilon_{Z\lambda} \, L \, [Z] + \varepsilon_{D\lambda} \, L \, [D] = OD_\lambda - \varepsilon_{N\lambda} \, L \, [N] \tag{16}$$

where L is pathlength and OD is UV absorbance. The effect of light scattering on OD_λ was accounted for by using equation (15). Finally, an overall protein balance was used to check the consistency of the calculations.

Results and Discussion

Refolding model test.   To determine if the proposed refolding kinetic model accurately describes the renaturation of hCAB, dilution refolding experiments were performed and their results compared with model predicted data. Predictions were obtained by solving equations (3) to (7) and the equations in Table 1 using a Runge-Kutta algorithm. Figures 4 and 5 show a comparison between experimental and model calculated data at two representative refolding conditions. Other conditions investigated also showed very good agreement between experimental and predicted data, indicating that the proposed refolding kinetic model (Equations (2) to (14)) adequately describes hCAB renaturation.

Refolding yield optimization.   The goal of the optimization was to determine the temperature, pH, diafiltration rate, and final urea concentration which maximize the ratio [N]/[P], or the ratio of renatured biologically active protein to total protein in the system.   The optimization was done in two steps.   The first step involved inputting final urea concentrations (0.1M - 1M urea) and determining the optimal temperature, pH, and diafiltration rate at which to operate in order to maximize the final ratio of [N]/[P].   This optimization problem was solved using a quasi-Newton method and a finite difference gradient.   The optimization routine determined that the objective function {1-[N]/[P]} was minimized at a pH of 9.0 and a temperature of 23 °C for all diafiltration rates and final urea concentrations examined.   Values of the [N]/[P] for various diafiltration rates and final urea concentrations are presented in Figure 6.   It is

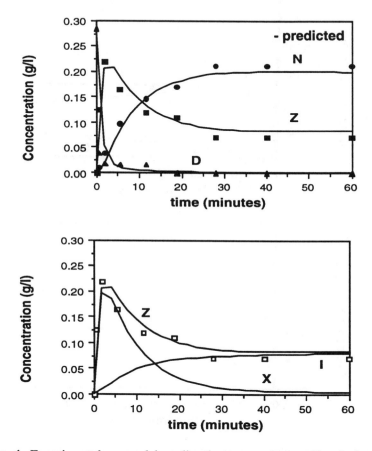

Figure 4. Experimental vs. model predicted concentration profiles during hCAB refolding at T = 23°C, pH = 8.85, [U] = 2.0 M, [hCAB] = 0.284 g/l

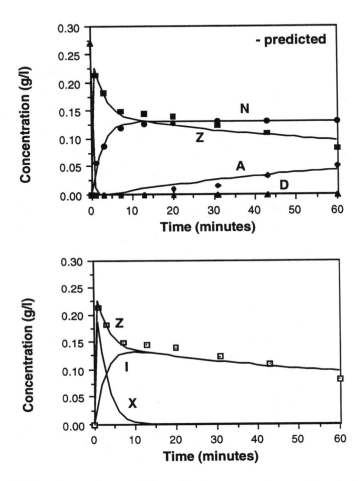

Figure 5. Experimental vs. model predicted concentration profiles during hCAB refolding at T = 37°C, pH = 7.45, [U] = 1.0 M, [hCAB] = 0.270 g/l

evident that optimal hCAB renaturation occurs at low final urea concentrations (<0.4M urea) at an F/V ratio circa 0.08-0.09 min^{-1}. Subsequently, using the previously determined optimal pH and temperature of 9.0 and 23 °C respectively, the optimal diafiltration profile and final urea concentration were evaluated using a procedure similar to that used in step one. The optimal diafiltration rate was determined to be 0.088 min^{-1} and no added benefit was obtained diafiltering to a final urea concentration of less than 0.3M urea.

The mathematically optimized hCAB renaturation conditions were tested in a laboratory scale diafiltration system. Diafiltration rates were adjusted by measuring filtration rates every 2-3 minutes and manually adjusting transmembrane pressure with a valve on the retentate line. The time necessary to reach the desired final urea concentration was previously calculated and the diafiltration was terminated at this time. The return of biologically active protein (in quadruplicate) as well as the UV/vis absorbance spectrum of the sample was immediately determined. An hour after completion of the diafiltration, hCAB reactivation was reevaluated and in all cases was found to be the same as previously measured. A comparison of experimentally determined renaturation yields with model predicted yields for the optimal conditions is presented in Table 3. The data demonstrate the predictive value of the approach to optimizing protein refolding yields described in this study. For the optimal F/V ratio of 0.088 min^{-1}, the mathematical model predicted a 73% renaturation yield, while the average renaturation yield from three diafiltration experiments under the conditions described was calculated to be 69%. Model predicted concentration profiles during diafiltration at optimal conditions are presented in Figure 7.

Table 3: **Optimization data at 23°C, pH = 9.0, $[U]_{final}$ = 0.3 M and F/V = 0.088 min^{-1}**

	% N	% D	% X	% I	% A
Experimental	69	~ 0	~ 0	26-31	0 - 5
Calculated	73	~ 0	~ 0	27	~ 0

The results of this study show the predictive value of this approach and its potential for optimizing protein refolding yields in general. It should be recognized that while diafiltration was selected as the means of denaturing agent removal in this study, this approach is also compatible with refolding by controlled dilution. It should also be recognized that this approach will potentially be most effective in optimizing renaturation yields of proteins which benefit from incubation in intermediate denaturant concentrations prior to final dilution. Builder and Ogez (*13*) have discussed several proteins, including urokinase, porcine growth hormone, and a foot and mouth virus capsid protein, for which refolding could potentially be optimized using the approach outlined in this study.

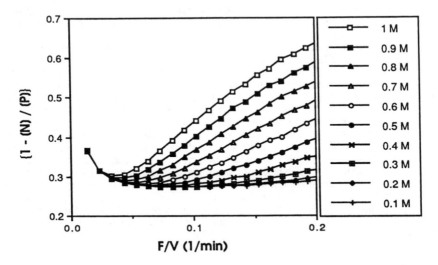

Figure 6. Effect of urea concentration and diafiltration rate on the objective function (equation 11) of the optimization problem at T = 23°C, pH = 9.0

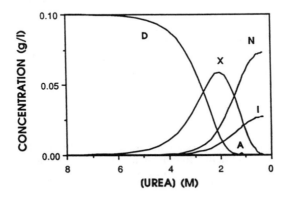

Figure 7. Predicted concentration profiles during ultrafiltration at optimal conditions, T = 23°C, pH = 9.0, F/V = 0.088 min^{-1}

Acknowledgments

The authors would like to thank Dr. Randall Swartz, Biotechnology Engineering Center, Tufts University for helpful discussions and the Amicon Corporation for providing ultrafiltration equipment.

Literature Cited

1. Kane, J.F. and Hartley, D.L. *Trends in Biotechnology* **1988**, *6*, 95-101.
2. Marston, F.A.O. *Biochem. J.* **1986**, *240*, 1-12.
3. Knuth, M.W. and Burgess, R.R. In *Protein Purification: Micro to Macro*, R. Burgess, ed., Alan R. Liss, Inc. **1987**.
4. Lowe, P.A., Rhind, S.K., Sugrue, R., and Marston, F.A.O. In *Protein Purification: Micro to Macro*, R. Burgess, ed., Alan R. Liss, Inc. **1987**.
5. Baldwin, R.L. and Eisenberg, D. In *Protein Engineering*, Alan R. Liss, Inc., **1987**, 127-148.
6. Ghelis, C. and Yon, J. In *Protein Folding*, Academic Press, NY, NY **1982**, 220-373.
7. Jaenicke, R. and Rudolph, R. In *Biological Oxidations,* H. Sund and V. Ullrich, edts. Springer Verlag, Heidelberg, Germany **1983**.
8. Mozhaev, V.V. and Martinek, K. *Enzyme Microb. Technol.* **1982**, *4*, 299-309.
9. Zettlmeissl, G., Rudolph, R., and Jaenicke, R. *Biochemistry* **1979**, *18*, 5567-5571.
10. London, J., Skrzynia, C., and Goldberg, M.E. *Eur. J. Biochem.* **1974**, *47*, 409-415.
11. Carlsson, U., Aasa, R., Henderson, L.E., Jonsson, B.H., and Lindskog,S. *Eur. J. Biochem.* **1975**, *52*, 25-36.
12. Orsini, G. and Goldberg, M.E. *The Journal of Biological Chemistry* **1978**, *253*, 3453-3458.
13. Builder, S. E. and Ogez, J.R. *U.S. Patent 4,620,948* **1986**
14. Ikai, A., Tanaka, S., and Noda, H. *Arch. Biochem. Biophys.* **1978**, *190*, 39-45.
15. Stein, P.J. and Henkens, R.W. *J. Biol. Chem.* **1978**, *253*, 8016-8018.
16. Jagannadham, M.V. and Balasubramanian, D. *FEBS Letters* **1985**, *188*, 326-330.
17. Semisotnov, G.V., Rodionova, N.A., Kutyshenko, V.P., Ebert, B., Blanck, J., and Ptitsyn, O. *FEBS Letters* **1987**, *224*, 9-13.
18. Vicik, S.M. Ph.D. Thesis, Tufts University, Department of Chemical Engineering **1990**.
19. Tanford, C., Aune, K.C. and Ikai. A. *J. Mol. Biol.* **1973**, *73*, 185-197.
20. Epstein, F.E., Schechter, A.N., Chen, R.F., and Anfinsen, C.B. *J. Mol. Biol.* **1971**, *60*, 499-508.
21. Kita, N., Kuwajima, K., Nitta, K., and Sugai, S. (1976). *Biochimica et Biophysica Acta* **1976**, *427*, 350-358.
22. Bower, V.E. and Robinson, R.A. *J. Phys. Chem.* **1963**, *67*, 1524-1527.
23. Pocker, Y. and Stone, J.T. *J. Am. Chem. Soc.* **1965**, *87*, 5497-5498.
24. Pocker, Y. and Stone, J.T. *Biochem.* **1967**, *6*, 668-678.
25. Creighton, T.E. *Methods in Enzymology* **1986**, *131*, 156-172.
26. Edsall, J.T., Mehta, S., Myers, D.V., and Armstrong, J.M. *Biochemische Zeitschrift* **1966**, *345*, 9-36.

27. Leach, S.J. and Scheraga, H.A. *J. Am. Chem. Soc.* **1960**, *82*, 4790-4792.
28. Garcia-Rubio, L.H. *Chem. Eng. Comm.* **1989**, *80*, 193-210.
29. Ko, B.P.N., Yazgan, A., Yeagle, P.L., Lottich, S.C., and Henkens, R.W. *Biochemistry* **1977**, *16*, 1720-1725.
30. Creighton, T.E. *Adv. Biophys.* **1974**, *18*, 1-20.
31. Dubois, T., Guillard, R., Prieels, J.P., and Perraudin, J.P. *Biochemistry* **1982**, *21*, 6516-6523.
32. Andersson, B., Nyman, P.O., and Strid, L. *Biochem. Biophys. Res. Comm.* **1972**, *48*, 670-677.
33. Parente, E.S. and Wetlaufer, D.B. *J. Chromat.* **1984**, *288*, 389-398.
34. Withka, J., Moncuse, P., Baziotis, A., and Maskiewicz, R. *J. Chromat.*, **1987**, *398*, 175-202.
35. Wong, K. and Tanford, C. *J. Biol. Chem.* **1973**, *248*, 8518-8523.
36. Carlsson, U., Henderson, L.E., and Lindskog, S. *Biochim. Biophys Acta* **1973**, *310*, 376-387.
37. Henkens, R.W., Kitchell, B.B., Lottich, S.C., Stein, P.J., and Williams, T.J. *Biochemistry* **1982**, *21*, 5918-5923.
38. Dolgikh, D.A., Kolomiets, A.P., Bolotina, I.A., and Ptitsyn, O.B. *FEBS Letters* **1984**, *168*, 331-334

RECEIVED March 22, 1991

Chapter 15

Commercial-Scale Refolding of Recombinant Methionyl Bovine Somatotropin

S. Bradley Storrs and Todd M. Przybycien[1]

Animal Sciences Division, Monsanto Agricultural Company, 700 Chesterfield Village Parkway, St. Louis, MO 63198

A method to dissolve and oxidize recombinant methionyl bovine somatotropin (mBST) from bacterial inclusion bodies suitable for large scale manufacture has been developed. mBST-containing inclusion bodies are dissolved, refolded, and oxidized in a single step in 4.5 M urea and pH 11.25 at protein concentrations in the 5 – 15 g/L range with yields in excess of 80%. Yield is a strong function of the urea concentration used during the oxidation reaction; optimal yields are obtained at intermediate urea concentrations.

Recombinant proteins expressed at high levels in bacterial hosts are often found in the form of inclusion bodies (1). These inclusion bodies consist of dense masses of partially folded, reduced protein. In this state, the target proteins are inactive; the inclusion bodies must be dissolved and the soluble protein must be refolded and oxidized into the native, active state. The typical downstream process for recovering protein from inclusion bodies includes two distinct operations: the dissolution of the inclusion bodies at high concentrations of denaturant such as urea or guanidine hydrochloride followed by a dilution or gradual removal of the denaturant to permit folding and oxidation to occur (2–4).

In this paper, we describe a one-step technique used at commercial scales for the dissolution, refolding, and oxidation of recombinant methionyl bovine somatotropin (mBST) from *E. coli* inclusion bodies (5). This technique employs moderate levels of a denaturant, urea, capable of disrupting the non-specific protein-protein interactions between mBST molecules within inclusion bodies. At these urea concentrations, soluble reduced mBST attains a partially folded

[1]Current address: Rensselaer Polytechnic Institute, Department of Chemical Engineering, Troy, NY 12180–3590

conformation (*Storrs and co-workers unpublished results, 6,7*) that may be readily refolded and oxidized (renatured) to the native state. This process results in a simplification of the traditionally used refolding protocols and may be more widely applicable to the production of other recombinant protein products that have intermediate or "molten globule" conformations in the reduced state (*8*).

Materials and Methods

Materials. Inclusion bodies containing mBST were isolated from recombinant *E. coli* similarly to that described by Bogosian and co-workers (*9*). The inclusion body preparation was stored frozen and thawed just prior to use.

Stock urea solutions were prepared from either ultra-pure urea purchased from Schwarz/Mann Biotech or urea purchased from Fisher Scientific that was further purified by deionization. Deionization was performed by passage of urea solutions through a high capacity mixed-bed ion-exchange cartridge (Barnstead number D8901) followed by an ultra-pure mixed-bed ion-exchange cartridge (Barnstead number D8902). Deionized water was used throughout. All other reagents were analytical grade or better.

Methods. The recovery of oxidized mBST monomer from inclusion body preparations involves the dissolution of the inclusion bodies and the subsequent air oxidation and refolding of the soluble reduced protein. Protocols for the investigation of these processes are described below. All experiments were performed at 5°C to minimize mBST degradation unless otherwise noted.

Quantitation of mBST and Determination of Monomer Yield. Oxidized monomeric mBST and total reduced mBST concentrations were measured by reverse-phase high pressure liquid chromatography (HPLC). Reduced and oxidized mBST and other inclusion body components were resolved on a Vydac C-18 column using a 50–60% acetonitrile gradient containing 0.1% trifluoroacetic acid at room temperature. Samples of oxidized mBST were prepared for analysis by dilution to approximately 1 mg/mL with a 4.5 M urea solution in 50 mM Tris buffer, pH 11. Samples for the determination of the total reduced mBST concentration of inclusion body suspensions and oxidized urea solutions were prepared by diluting aliquots to approximately 1 mg/mL with a solution containing 5% sodium dodecyl sulfate and 100 mM dithiothreitol. Quantitation of total reduced mBST in inclusion body suspensions prior to dissolution or in the final urea solution gave similar values. Yields of oxidized monomeric mBST were calculated by dividing the oxidized monomer concentration by the total reduced mBST concentration.

Dissolution of mBST Inclusion Bodies. The dissolution solutions were prepared by blending inclusion bodies, 8 M urea stock solution, 2 M Tris-Cl solution at the target pH, and water to give a final mBST concentration of 2 mg/mL at the desired pH and urea concentration and a total volume of 1.5 L. The solutions were magnetically stirred for 1–2 hours and then clarified by centrifugation at 12,000 ×g for 15 minutes and passage through a 0.2 μm syringe filter. The mBST concentration of the filtrate was determined by the HPLC technique described above.

Determination of the Urea Dependence of Renaturation Yields. Renaturation solutions were prepared by mixing inclusion bodies, 7.5 M urea stock solution, and water to a final mBST concentration of 9.5 mg/mL at the desired urea concentration and a total volume of 100 mL. The solution pH was adjusted to 11 with 2.5 M NaOH and magnetically stirred in an open container to promote the air oxidation of the reduced mBST. When the oxidation was complete, usually after 2 days as judged by reverse-phase HPLC, the samples were analyzed for monomeric mBST. Samples oxidized at urea concentrations below 4 M were either dissolved at higher pH and adjusted back to pH 11 or dissolved at higher urea concentration and diluted back to the desired concentration.

Determination of the pH Dependence of Renaturation Yields. Renaturation solutions were prepared by mixing inclusion bodies, 7.5 M urea stock solution, and water to final concentrations of 4.5 M urea and 8 mg/mL mBST with a final volume of 25 mL. Samples renatured at pH 11.5 and above were dissolved at the same pH; samples renatured below pH 11.5 were first dissolved at pH 11.5 and then adjusted back to the desired pH with dilute HCl. Samples were taken for monomeric mBST analysis following air oxidation.

Determination of the Protein Concentration Dependence of Renaturation Yields. Inclusion bodies, 7.5 M urea stock solution, and water were mixed to give a final urea concentration of 4.5 M and the desired protein concentration at a final volume of 25 mL. Following pH adjustment to 11 with NaOH, the solutions were air oxidized and analyzed as above.

Results and Discussion

Dissolution of Inclusion Bodies. The dissolution behavior of mBST containing inclusion bodies as a function of urea concentration and pH is shown in Figure 1. As the pH of the dissolution solution is increased from 8 to 11, the urea concentration required for complete dissolution of mBST inclusion body suspensions decreased significantly. At pH 8, complete dissolution necessitated the use of 7.5 M urea concentrations; at pH 11, 4.5 M urea was sufficient. This pH effect is likely due to the poor solubility of mBST near neutral pHs. The use of alkaline pHs for inclusion body dissolution allows substantial savings, in terms of the decreased urea raw material costs and the concomitant reduction in urea waste, to be realized. Since intermediate urea concentrations at alkaline pHs were used successfully in the subsequent renaturation process, the dissolution and renaturation processes could be accomplished in a single step.

Oxidation of mBST in Urea Solutions. Three parameters were found to strongly influence monomer yield during the air oxidation of mBST in urea solution: urea concentration, pH, and mBST concentration. Yield is defined as the final concentration of recovered oxidized monomeric mBST relative to the total initial mBST concentration as measured by the HPLC technique described in the Materials and Methods section. Size exclusion HPLC analyses of final renatured mBST solutions and corresponding chemically reduced samples (data not shown) indicate that yield losses are primarily due to the formation of disulfide-linked intermolecular aggregates.

Urea Concentration Dependence. The effect of urea concentration over the 1 to 7 M range on renaturation yields at pH 11 and an mBST concentration of 9.5 mg/mL is shown in Figure 2. Because of the marked decrease in mBST inclusion body solubility at urea concetrations below 4 M at pH 11, two modified procedures were used to dissolve the inclusion bodies prior to renaturation. In the first procedure, inclusion bodies were dissolved at 9.5 mg/mL at elevated pH (11.8–12.2) in 1, 2, and 3 M urea solutions and upon completion of dissolution, the pH was adjusted to 11 with HCl. The alternative procedure involved solubilizing higher concentrations of inclusion bodies in 4.5 M urea at pH 11 followed by dilution of the solutions with 10^{-3} M NaOH to an mBST concentration of 9.5 mg/mL and the appropriate urea concentration. Similar renaturation yield results were obtained via both techniques as indicated in Figure 2.

Optimal renaturation yields were obtained at intermediate urea concentrations; a maximum yield of 66% was found in 4.5 M urea in this series of experiments. This urea concentration dependence may be due to a balance between aggregative and denaturing forces dominant at low and high urea concentrations, respectively.

pH dependence. The effect of pH in the 9 to 12.5 range on mBST renaturation yields at an mBST concentration of 4.5 M urea is given in Figure 3. As above, the reduced solubility of mBST inclusion bodies in 4.5 M urea below pH 11 necessitated a modification of the procedure for experiments targeted for the 9 to 10.5 pH range. For these experiements, mBST inclusion bodies were dissolved at 8 mg/mL in 4.5 M urea at pH 11 and then the pH was adjusted to the desired point with HCl. Data was not collected below pH 9 since reduced mBST in 4.5 M urea precipitates. The optimal renaturation yields occurred at

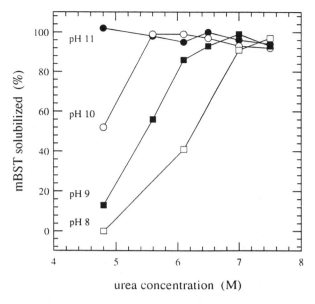

Figure 1. Dissolution of mBST inclusion bodies as a function of urea concentration and pH. Total protein concentration was 2 mg/mL.

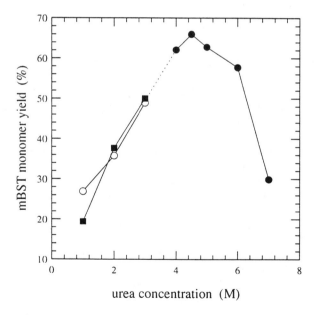

Figure 2. mBST renaturation yield as a function of urea concentration. pH was constant at 11 and mBST concentration was 9.5 mg/mL. Symbols: inclusion body dissolution at pH 11 (●), high pH (■), and high urea concentration (○).

pHs of 11–11.5. The rapid decline in yield above pH 11.5 is possibly due to the reduction of repulsive protein-protein interactions as the lysine and arginine residues are titrated, leading to increased aggregation. Below pH 11, the decline may be related to the decreasing solubility of reduced mBST.

Protein Concentration Dependence. The oxidation and refolding of mBST in 4.5 M urea at pH 11 was strongly dependent on the concentration of mBST during renaturation. Figure 4 shows renaturation yields for mBST concentrations between 1 and 30 mg/mL. The effect of mBST concentration on monomer yield was relatively linear with a slope of approximately 1.5% yield loss per mg/mL increase in mBST concentration. Monomer yields of 80–85% were obtained at mBST concentrations of 5 mg/mL. It is likely that the increase in yield losses at increasing mBST concentrations is due to an increase in the kinetic rates of nonspecific aggregation reactions.

The conditions for maximal mBST renaturation yield in urea solution were found to be inclusion body dissolution in 4.5 M urea at pH 11–11.5 at total protein concentrations of 5–15 mg/mL followed by air oxidation over several days. At pH 11-11.5 the protein itself has substantial buffering capacity and no additional bufferring agents are necessary to maintain a constant pH. Therefore, the only materials required for this process are inclusion bodies, urea, and NaOH.

Conclusions

A method to dissolve and oxidize recombinant methionyl bovine somatotropin from bacterial inclusion bodies in a single step suitable for large-scale manufacture has been developed (5). mBST inclusion bodies are solubilized and the soluble mBST is renatured (refolded and air-oxidized) in 4.5 M urea at pH 11-11.5. This process is performed over the course of several days at 5°C with total protein concentrations of 5-15 mg/mL. The urea used in this process is removed in subsequent steps via conventional techniques. The procedure is a simplification of traditional two-step operations and results in high active monomer yields at high protein concentrations. The performance of this process can be described in terms of recent findings in protein folding research.

Yield is a strong function of the urea concentration used during the oxidation reaction with optimal yields occuring in the vicinity of 4.5 M urea. We propose that this optimum is the result of a balance between aggregative processes at low urea concentrations and unfolding processes at high urea concentrations. The tradeoff between aggregation and unfolding has been described in the optimal renaturation of reduced chymotrypsinogen A with low concentrations of guanidine hydrochloride (10). The existence of optimum denaturant concentrations with respect to native protein yields has also been noted for the renaturation of human carbonic anhydrase (11) and bovine serum albumin (12).

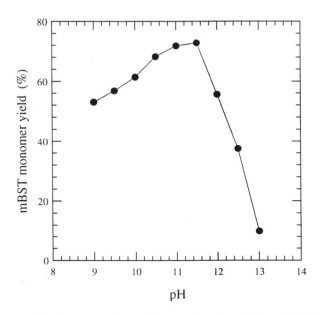

Figure 3. mBST renaturation yield as a function of pH at 4.5 M urea and 8 mg/mL mBST. See text for sample preparation details.

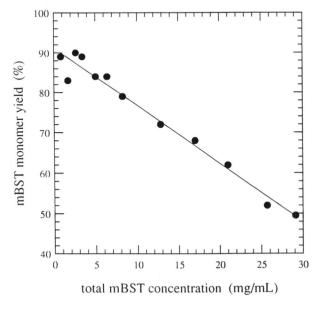

Figure 4. mBST renaturation yield as a function of protein concentration at pH 11 and 4.5 M urea. Line is least squares fit to experimental data.

The key to this process appears to be the existence of a molten globule folding intermediate for mBST that is maximally populated at moderate denaturant concentrations. Intermediate denaturant levels allow the efficient release of protein from inclusion bodies. The molten globule state is typically native-like in secondary structure content and has a compact tertiary structure $(13,14)$; this intermediate state is thought to be "poised" for correct refolding and oxidation into the native conformation $(8,15)$. In this state, existing secondary structure elements likely serve to restrict conformational degress of freedom and to preposition sulfhydryl moeities for correct disulfide bond formation. This has been termed the "framework model" for protein folding $(7,14,15)$.

This renaturation procedure may find wider application to the commercial-scale renaturation of other recombinant proteins from bacterial inclusion bodies. The only proviso appears to be the existence of one or possibly more partially folded states at intermediate denaturant concentrations. Structural homologs of BST such as porcine somatotropin (16) and bovine prolactin are candidates for this approach to refolding. We are currently developing a general model for the prediction of optimum denaturant concentration for protein refolding via a consideration of the tradeoffs between aggregation and unfolding phenomena.

Literature Cited

1. Krueger, J.K.; Stock, A.M.; Schutt, C.E.; Stock, J.B.; In *Protein Folding: Deciphering the Second Half of the Genetic Code*; Gierasch, L.M.; King, J. Eds.; American Association for the Advancement of Science: Washington, D.C., 1990; pp 136-142.

2. Schein, C.H. *Bio/Technol.* **1990**, *8*, 308-316.

3. Ogez, J.R.; Hodgdon, J.C.; Beal, M.P.; Builder, S.E. *Biotech. Adv.* **1989**, *7*, 467-488.

4. Marston, F.A.O. *Biochem. J.,* **1986**, *1*, 1-12.

5. Bentle, L.A.; Mitchell, J.W.; Storrs, S.B. *U.S. Patent 4652630* **1987**.

6. Brems, D.N.; In *Protein Folding: Deciphering the Second Half of the Genetic Code*; Gierasch, L.M., and King, J. Eds.; American Association for the Advancement of Science: Washington, D.C., 1990; pp 129-135.

7. Holzman, T.F.; Brems, D.N.; Dougherty, J.J. *Biochemistry* **1986**, *25*, 6907-6917.

8. Ptitsyn, O.B.; Pain, R.H.; Semisotnov, G.V.; Zerovnik, E.; Razgulyaev *FEBS Lett.* **1990**, *262*, 20-24.

9. Bogosian, G.; Violand, B.N.; Dorward-King, E.J.; Workman, W.E.; Jung, P.E.; Kane, J.F. *J. Biol. Chem.* **1989**, *264*, 531-539.

10. Orsini, G.; Goldberg, M.E. *J. Biol. Chem.* **1978**, *253*, 3453-3458.

11. Carlsson, U.; Henderson, L.E.; Lindskog, S. *Biochim. Biophys. Acta* **1973**, *310*, 376-387.

12. Damodaran, S. *Biochim. Biophys. Acta* **1987**, *914*, 114-121.
13. Fischer, G.; Schmid, F.X. *Biochemistry* **1990**, *29*, 2205-2212.
14. Ikeguchi, M.; Sugai, S. *Int. J. Peptide Protein Res.* **1989**, *33*, 289-297.
15. Holzman, T.F.; Dougherty, J.J.; Brems, D.N.; MacKenzie, N.E. *Biochemistry* **1990**, *29*, 1255-1261.
16. Abdel-Meguid, S.S.; Shieh, H.-S.; Smith, W.W.; Dayringer, H.E.; Violand, B.N.; Bentle, L.A. *Proc. Natl. Acad. Sci. US* **1987**, *84*, 6434-6437.

RECEIVED March 22, 1991

Chapter 16

Large-Scale Refolding of Secretory Leukocyte Protease Inhibitor

Robert J. Seely and Mark D. Young

Synergen, Inc., Boulder, CO 80301

A lab-scale procedure for refolding the recombinant protein, secretory leukocyte protease inhibitor, was scaled to 1000 liter production batches. Optimization of reaction conditions by a statistical experimental design approach resulted in consistent activity recoveries of 80-85%, and lowered cost. The statistical design method allows simultaneous optimization of interacting process variables. Changes in the refold reaction conditions greatly influence the level of specific contaminants, thus purity becomes an important parameter in addition to yield. Our experience in the development, scale-up, and cost analysis of a protein refolding operation is presented.

Secretory leukocyte protease inhibitor is a protein of 11,726 molecular weight, secreted by the parotid gland. First isolated and characterized by Thompson and Ohlsson (1) in 1986, it was found to have a high affinity for leukocyte elastase, cathepsin G and trypsin. The inhibitor is of therapeutic interest in the treatment of disease states that involve leukocyte-mediated proteolysis. Emphysema and cystic fibrosis are examples where severe tissue destruction is caused by uncontrolled proteolytic activity.

The gene was cloned (2, 3) and expressed in a JM 109 strain of E. coli (4). The protein is accumulated intracellularly in an inactive form. Upon cell disruption by high pressure homogenization, recombinant secretory leukocyte protease inhibitor (rSLPI) is only partially soluble, although discrete inclusion bodies have not been identified. Having 16 cysteines involved in 8 intramolecular disulfides, the protein must be refolded to regain full stability and activity.

In this manuscript we describe our experience in the development and scale-up of the refolding procedure from initial lab preparations to a production scale of 175 g rSLPI in a 1000 liter refold batch. Optimization studies to

0097–6156/91/0470–0206$06.00/0

increase the yield and minimize cost and specific contaminants were performed before and during scale-up, and again during an inter-campaign period. In addition to straightforward reagent modifications, a statistical experimental design approach was employed to examine the chemical interactions of such refolding parameters as reducing and oxidizing agents, pH, time and protein concentration. The final operating conditions resulted in yields of 80-85%, at 0.2 g rSLPI per liter, in 4 hours.

Methods and Materials

Reversed-phase HPLC was performed on a SynChropak RP-8 (SynChrom, Inc., Linden, IN). The gradient was from 19% to 34% acetonitrile, at 1%/min., in 0.1% trifluoroacetic acid. This method resolves the correctly refolded rSLPI from the unfolded form.

Activity was determined by the ability of rSLPI to inhibit a known amount of trypsin, in a chromogenic assay using Tosyl-gly-pro-lys-4-nitranilide acetate (Chromozym PL, Boehringer-Mannheim).

A stock solution of trypsin (Sigma T-8642 or equivalent) is first standardized using the substrate p-nitrophenol-p'-guanidobenzoate HCl (Sigma N-8010). By the initial burst principle (5) one can titrate the active site of trypsin and calculate the uM amount of active trypsin introduced into the rSLPI inhibition assay. By knowing the molecular weight of the trypsin (source dependent) the uM amount of rSLPI present can be deduced from the 1:1 stoichiometry of the reaction.

All buffer components and refold reaction chemicals were USP grade.

The statistical experimental designs were central composite Box-Behnken models, incorporated into the software package, X-Stat (6).

Bench-Scale Refold Process

Initial refolding from a solution containing reduced, denatured rSLPI was established by a variety of workers at Synergen (see Acknowledgments). The basic approach was to ensure that the protein was fully reduced and denatured, then induce folding by dilution, away from the denaturant and into a cystine containing solution, which allows disulfide bond formation. The quantities of reagents required to accomplish this were determined empirically. The feedstream at that time was generated as a partially purified cell lysate containing 50 mM 2-mercaptoethanol (BME) and 5 mM ethylenediamine tetraacetic acid in a Tris(hydroxymethyl)aminomethane buffer (Tris) at pH 7.8.

The procedure was as follows: the feedstream was made 3 M in guanidine-HCl, allowed to incubate at room temperature for 30 minutes, then dithiothreitol (DTT) was added to 5 mM and the incubation continued for one hour. Cystine was added, in 0.5 M NaOH, to 14 mM, and mixed for 10 minutes. The mixture was then diluted into 10 volumes of a 50 mM Tris buffer, pH 8.0, containing 5.3 mM cysteine. The reaction was allowed to proceed at room temperature for 16 hours (7).

From the feedstream, 50-65% of the rSLPI could be refolded to the active species, at a final concentration of 0.05-0.1 g/l. The yield, quality, and ease of operation were adequate for small-scale production involving several subsequent chromatography steps. These refold conditions were thus successfully incorporated into the first pilot-scale process for production of gram amounts of rSLPI for preliminary animal testing studies. Figure 1 is a typical HPLC chromatogram of the refold mixture. Work in our laboratories has shown that reversed phase HPLC is a good method for following the course of the refolding reaction. Active rSLPI is at 13 min. retention, while the fully unfolded protein is at 18-19 min. and the partial or incorrectly folded species are found in between.

The results of the preliminary animal testing studies with the recombinant protein were positive and it became clear that much more material would be required for continued animal studies, and that the quality of the purified protein would be very important. The increase in scale of production necessitated a development program to ensure process reliability and quality. One result of this program led to the finding that the rSLPI as supplied to the refold step did not require additional denaturation, eliminating the requirement for guanidine. Upstream operations involve denaturation of rSLPI with 4 M urea and BME. Apparently the protein remains fully denatured and reduced through the initial isolation steps. This change resulted in a reproducible process for folding the protein in 100-500 liter (final volume) batches with an average refolding yield of 64% and an active rSLPI concentration of 0.1 g/l. The result of this work was that the scalability of the refold process was clearly demonstrated.

In addition to the consistent yield of rSLPI, there did not appear to be any new contaminants generated that could not be removed in subsequent purification steps, and there was no significant increase in breakdown products of the target protein.

Production of Clinical Lots

Once again, additional animal studies indicated that rSLPI held promise as a useful therapeutic agent, and a commitment to carry the product into formal toxicology and human clinical trials was made. This commitment necessitated the manufacture of kilogram quantities of the recombinant protein. Since the protein would be used in human clinical trials, an additional constraint was placed on the process: it must be manufactured in accordance with current Good Manufacturing Practices (cGMP) as prescribed by the FDA. Furthermore, process changes subsequent to this production campaign would have to be justified by extensive documentation of product quality and perhaps even additional clinical efficacy testing. This, then became the point at which considerable effort would be expended to develop a cost-effective, reproducible process for the manufacture of the bulk product.

The first series of experiments was designed to ask whether replacements for the buffer, Tris, and the reducing agent, DTT, could be found, since these were the two highest cost components of the refold protocol (see Cost of Refolding Proteins). Results at the bench showed that refold yield appeared to be unaffected by the elimination of DTT and Tris from the refold protocol. It was also found at this time that a variant form of rSLPI, which could be identified by reversed phase HPLC in the final product, might be formed in the refold step. This variant, called P-2, has been tentatively identified as an oxidation product of rSLPI. When the refold process was scaled, larger amounts of P-2 were found in the refold solutions, probably due to the increase in solution surface area exposed to air. Air oxidation of rSLPI to the P-2 species was verified at the small-scale, prompting the addition of an argon gas overlay during the refold step. The presence of argon throughout the refolding process greatly minimizes the formation of P-2. P-2 does not appear to be a rSLPI dimer; rather it can be generated by chemical oxidation. The generation of P-2 can represent a significant loss of rSLPI, and furthermore it presents a serious problem for purification.

Using the new refolding protocol, without DTT and Tris, the effects of cysteine and cystine levels on yield and production of the undesirable P-2 species were systematically investigated. It was found that the cystine level was an important determinant of yield and the cysteine level had a major impact on the formation of P-2 (Figure 2).

The refolding protocol developed from these studies was very simple: the partially purified rSLPI at 1.6 g/l, still in essentially the same milieu as for earlier work, is made 50 mM in L-cystine, then diluted with six volumes of purified water. L-cysteine HCl is added to a final concentration of 14 mM and the pH is adjusted to 8.2. The refold reaction is essentially complete in 4 hours at 18 to 20^0 C.

This process was scaled to provide refolded rSLPI in 1000 liter batches for downstream purification. The refold step was very reliable and reproducible, given a consistent feedstock of reduced rSLPI. Over the ensuing production campaign the process yielded an average refold recovery of 74%, at 0.15 g/l final concentration.

Post-Campaign Optimization

Process development activities to optimize rSLPI production for clinical trials were subject to time constraints. Subsequent to this production campaign, we are continuing to optimize the process for rSLPI, since it will be a commercially important product. It must, however, be recognized that process changes will be reviewed for their impact on the quality and efficacy of the final product.

The common features of protein folding that emerge from our studies and those in the literature are the importance of initial formation of mixed disulfides, which appear to promote intramolecular disulfides, and the continued presence of a reducing agent to allow disruption of any incorrect disulfide bonds that form during the process. In addition, our protein appears to require the reducing agent to prevent the formation of an oxidized variant. Chan (8) has reported on the necessity of a protonated amine on the mixed disulfide for

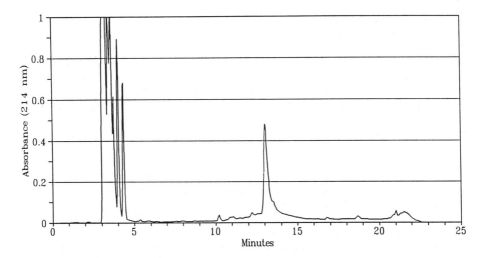

Figure 1. Reversed-phase HPLC chromatogram of a bench-scale refold mixture. The refold procedure was as described for the gram amount production lots for preliminary animal testing. Active rSLPI is at 13 min. retention, while the fully unfolded protein is at 18-19 min. and the partial or incorrectly folded species are found in between.

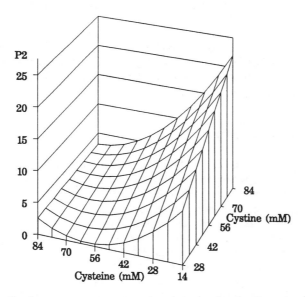

Figure 2. Surface response curve showing the level of production of the contaminant P-2, predicted by a statistically designed experimental matrix, as the relative levels of cystine (mM in the initial feedstream pool) and cysteine (mM in the final diluted refold mixture) are varied. P-2 is determined by reversed-phase HPLC, and is expressed as area percent of the chromatogram.

protein sulfonation. With cysteine, glutathione or mercaptoethanolamine, efficient nucleophilic attack by sulfite can occur and complete sulfonation of the protein is achieved. Mixed disulfides lacking a positive charge, such as mercaptoethanol or DTT, do not favor sulfonation. One can expect the formation of intramolecular disulfides to involve a similar mechanism, and indeed it appears that the former group of mixed disulfides do promote refolding, and they are commonly used by researchers (7). If this concept is correct one should also be able to replace cysteine with any other reducing agent. Lab studies have verified this; with BME as a direct replacement for cysteine, the reaction proceeds at the same rate and to the same extent.

With this change in chemical composition, a series of statistically designed experiments was performed. Variables of interest were identified as reaction time and the concentrations of rSLPI, cystine and BME. Using the Box-Behnken statistical design, a series of 27 experiments was done. Activity assays and reversed phase HPLC analyses generated data on yield of active rSLPI, relative purity, and relative levels of specific contaminants of interest. The results were modeled with quadratic equations by the X-Stat program and projections of maximum yield and purity, and minimum production of the contaminants, were obtained.

The relationship of yield to the concentrations of the sulfhydryl reagents is particularly dramatic. Figure 3 shows a pair of surface response curves for yield as generated by the quadratic model. Figure 3a displays the effects of cystine and rSLPI concentration on yield. Clearly, higher levels of cystine are beneficial. The interrelation of cystine and rSLPI concentration is also demonstrated. As levels of rSLPI are increased, the need for increased cystine is apparent but yield nevertheless declines with increasing rSLPI concentration, presumably for some other reason. Figure 3b diagrams the interaction of cystine and BME; high levels of both have a marked influence on yield. Reaction time did not show a significant interaction with the other variables (these results are not shown).

The experimental conditions that provide for optimization of one parameter often conflict with the optimization of another. This is shown in Figure 4. The high level of cystine that promotes yield also gives rise to the contaminant P-2. High BME concentration does help to suppress this, however. Interactive responses such as this can offer significant information in understanding possible mechanisms, and we are continuing to study the possibility that P-2 is an oxidative derivative of rSLPI.

For maximum yield the model specifies the following conditions: cystine at 50 mM, BME at 20 mM, and a 4 hour incubation with the final concentration of rSLPI at 0.15 g/l. These conditions compromise relative purity and contaminant production (shown by the same model), and thus were adjusted slightly with these considerations in mind. A test refold with conditions near those specified by the model gave a yield of 83%. The predicted value under the chosen conditions was 85%. As seen in Figure 3, the levels of BME and cystine fall short of creating an optimum in yield; thus the ranges could be extended further. These studies are underway at this time. We are also

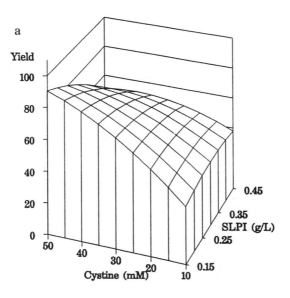

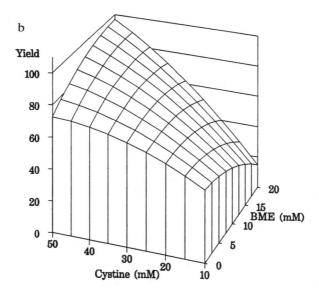

Figure 3. Surface response curves showing the interactive effects of cystine and total rSLPI concentration on the percent yield of active rSLPI (3a). 3b represents the effects of cystine and BME on the percent yield of active rSLPI.

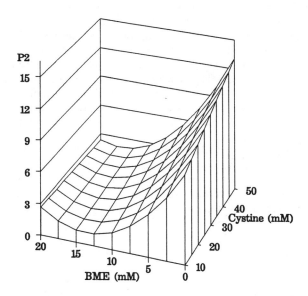

Figure 4. Surface response curves for contaminant P-2 as a function of cystine and BME concentration. P-2 is expressed as in Figure 2.

modeling optimization objective functions that are dependent on yield, but incorporate a penalty for contamination production.

Cost of Refolding Proteins

The ultimate goal of protein refolding studies in the process development environment is to develop a cost effective process that is simple and scalable. Our approach to the development of the overall process, including the refold step, is to use cost models to evaluate and guide the development activity.

There are two important factors in determining the cost of the refold step. The first is the cost of labor, utilities, equipment, and raw materials to produce the refolded protein. The second is the purifiability of the protein from the other components of the environment in which it is refolded. This is not a precisely definable function of the measurable properties of the refolded protein, but must be evaluated by its downstream performance in the process with respect to yield and quality. A further complication arises in that the process may not be fully developed and therefore changes in the process will occur that will improve both yield and quality, independent of the refold conditions. Bearing these conditions in mind, we present cost information in this section on the basis of the cost to refold the protein. Costs associated with its purifiability are not presented.

Using the simple liquid batch reaction approach to the refold step, our studies have shown that the only significant cost variable, when the process is projected to large-scale, is the cost of the raw materials. For the purposes of the studies reported here, we have used the costs reported in Table I. These are our current costs for materials of appropriate purity for this purpose.

Table I. Unit Costs of Raw Materials Used in Refolding

Raw Material	Unit Cost, $/gram
2-Mercaptoethanol	0.0275
L-Cystine	0.1095
L-Cysteine HCl	0.1435
Guanidine HCl	0.0088
Tris	0.0380
Dithiothreitol	7.0000

Cost of the Bench-scale Process. The cost of materials for the treatment of 1 liter of partially purified rSLPI, containing approximately 1 gram of the protein of interest was $20.65. At a refold yield of 50%, this represented a cost of refolding of $41.29 per active gram of rSLPI produced. Note that this cost calculation does not include the upstream cost of preparing the partially purified rSLPI.

Refolding Cost of the Clinical Production Campaign. The cost of materials for treating 1 liter of partially purified rSLPI, containing approximately 1.8 grams, was $3.76. The average refold yield for the five best runs of the campaign was 83%, resulting in a refold step cost of $4.53 per gram of active rSLPI produced.

Refolding Cost of Post-Campaign Process. The replacement of cysteine with BME in the refold will result in considerable reduction in cost of the process, since BME is approximately 11% the cost of cysteine on a molar basis.

Discussion

From our experience with rSLPI and other, more recalcitrant proteins, it appears that refolding and correct disulfide bond formation can be relatively rapid and complete in the presence of adequate disulfide catalysts (7). The major difficulty is not in providing a proper environment to promote refolding, but rather in maintaining solubility and preventing aggregation. Beyond these stability concerns, there is a second class of proteins for which activity is difficult to restore. For proteins that are co-translationally modified either by covalent alterations (9) or by chaperone interaction (10), it is extremely difficult to recreate the mammalian intracellular environment. Extensive genetic modification or expression in a eukaryotic cell may be the only alternatives.

Once solubility problems have been overcome, rapid optimization of refold yield can be achieved, provided an accurate assay for the correctly folded species has been developed. The complexity of the reaction is increased because many of the variables interact with each other. We have demonstrated the effectiveness of addressing the interaction of chemical-physical conditions by a statistical experimental design approach. This allows one to optimize a host of variables without performing experiments involving every single combination.

When acceptable yields and costs are achieved, scale-up is a minor barrier for solution-type refolding. One must, however, be concerned with subtle scale differences such as mixing or air-liquid surface area changes, and the consequent generation of new contaminants; either host cell proteins that are retained soluble, or breakdown products of the target protein itself.

A cost of production model serves to not only monitor the improvements in optimization in a meaningful way, but also to direct future research efforts toward cost sensitive areas or chemicals.

Acknowledgments

We wish to thank following coworkers for their significant contributions to this project: Tadahiko Kohno and Michael Betlach for bench-scale refolding technique development; Karin Hale and Robert Thompson for development of the active site titration method for assaying rSLPI; Holly Fry and Lea C. Baca for performing the multiple experiments and activity assays;

Randal Hassler and Michael Betlach for assistance in X-Stat use and graphic analysis.

Literature Cited

1. Thompson, R.C.; Ohlsson, K. *Proc. Natl. Acad. Sci., U.S.A.* **1986**, *83*, 6692-6696.
2. Stetler, G.L.; Brewer, M.T.; Thompson, R.C. *Nucleic Acids Res.* **1986**, *14*, 7883-7895.
3. Ohlsson, K.; Rosengren, M.; Stetler, G.; Brewer, M.; Hale, K.K.; Thompson, R.C. In *Pulmonary Emphysema and Proteolysis:1986*; Taylor, J.C. and Mittman, C., Eds.; Academic Press: Orlando, FL, **1987**, pp 307-324.
4. Eisenberg, S. P.; Hale, K. K.; Heimdal, P.; Thompson, R. C. *J. Biol. Chem.* **1990**, *265*, 7976-7981.
5. Fersht, A. *Enzyme Structure and Function*; W.H. Freeman & Co.: San Francisco, CA, **1977**; p 122.
6. Murray, J.S. *X-Stat Statistical Experimental Design, Data Analysis and Nonlinear Optimization*; John Wiley & Sons: New York, NY, **1985**.
7. Kohno, T.; Carmichael, D.F.; Sommer, A.; Thompson, R.C. *Methods in Enzymology*. **1990**, *185*, 187-195.
8. Chan, W. *Biochemistry*. **1968**, *7*, 4247-4254.
9. King, J. *Bio/Technol.* **1986**, *4*, 297-303.
10. Hemmingsen, S.M.; Woolford, C.; van der Vies, S.M.; Tilly, K.; Dennis, D.T.; Georgopoulos, C.P.; Hendrix, R.W.; Ellis, R.J. *Nature*. **1988**, *333*, 330-334.

RECEIVED February 6, 1991

Author Index

Affiliation Index

Subject Index

Production: Margaret J. Brown
Indexing: A. Maureen Rouhi
Acquisition: Barbara C. Tansill
Cover design: Tina Mion

Printed and bound by Maple Press, York, PA

Bestsellers from ACS Books

The ACS Style Guide: A Manual for Authors and Editors
Edited by Janet S. Dodd
264 pp; clothbound, ISBN 0–8412–0917–0; paperback, ISBN 0–8412–0943–X

Chemical Activities and Chemical Activities: Teacher Edition
By Christie L. Borgford and Lee R. Summerlin
330 pp; spiralbound, ISBN 0–8412–1417–4; teacher ed. ISBN 0–8412–1416–6

Chemical Demonstrations: A Sourcebook for Teachers,
Volumes 1 and 2, Second Edition
Volume 1 by Lee R. Summerlin and James L. Ealy, Jr.;
Vol. 1, 198 pp; spiralbound, ISBN 0–8412–1481–6;
Volume 2 by Lee R. Summerlin, Christie L. Borgford, and Julie B. Ealy
Vol. 2, 234 pp; spiralbound, ISBN 0–8412–1535–9

Writing the Laboratory Notebook
By Howard M. Kanare
145 pp; clothbound, ISBN 0–8412–0906–5; paperback, ISBN 0–8412–0933–2

Developing a Chemical Hygiene Plan
By Jay A. Young, Warren K. Kingsley, and George H. Wahl, Jr.
paperback, ISBN 0–8412–1876–5

Introduction to Microwave Sample Preparation: Theory and Practice
Edited by H. M. Kingston and Lois B. Jassie
263 pp; clothbound, ISBN 0–8412–1450–6

Principles of Environmental Sampling
Edited by Lawrence H. Keith
ACS Professional Reference Book; 458 pp;
clothbound; ISBN 0–8412–1173–6; paperback, ISBN 0–8412–1437–9

Biotechnology and Materials Science: Chemistry for the Future
Edited by Mary L. Good (Jacqueline K. Barton, Associate Editor)
135 pp; clothbound, ISBN 0–8412–1472–7; paperback, ISBN 0–8412–1473–5

Personal Computers for Scientists: A Byte at a Time
By Glenn I. Ouchi
276 pp; clothbound, ISBN 0–8412–1000–4; paperback, ISBN 0–8412–1001–2

Polymers in Aqueous Media: Performance Through Association
Edited by J. Edward Glass
Advances in Chemistry Series 223; 575 pp;
clothbound, ISBN 0–8412–1548–0

For further information and a free catalog of ACS books, contact:
American Chemical Society
Distribution Office, Department 225
1155 16th Street, NW, Washington, DC 20036
Telephone 800–227–5558

BC

OCT 14 1991